Honeyland

I0826278

The fourth volume in the Docalogue series, this book explores the significance of the documentary *Honeyland* (2019) in relation to documentary ethics, the representation of human and animal relations, environmental studies, genre theory, and documentary distribution.

The film, focused on a Turkish-speaking woman in Macedonia who cultivates bees to produce honey through an ancient and environmentally sustainable method, raises important questions about the place of humans and economic activity within the broader ecosystem. The documentary also prompts critical reflection about the relationship between observation and storytelling, how the film festival circuit allows certain films to reach a wide audience, the ethics of ethnographic representation, the relationship between human and insect life, and to what extent film can allow us to experience others' life-worlds. By combining five distinct critical perspectives on a single documentary, this book acts both as an intensive scholarly treatment of the film and as a guide for how to analyze, theorize, and contextualize a documentary text.

This book will be of interest to students and scholars of documentary studies, as well as those studying film and media more broadly.

Jaimie Baron is a Professor of Film Studies at the University of Alberta. She is the author of two books, *The Archive Effect: Found Footage and the Audiovisual Experience of History* (2014) and *Reuse, Misuse, Abuse: The Ethics of Audiovisual Appropriation in the Digital Era* (2020), and numerous journal articles and book chapters. She is also the director of the Festival of (In)appropriation, a yearly international festival of short experimental found footage films and videos.

Kristen Fuhs is an Associate Professor of Media Studies at Woodbury University. She writes about documentary film, the American criminal justice system, and contemporary celebrity, and her work has appeared in journals such as *Cultural Studies*, the *Historical Journal of Film, Radio, and Television*, and the *Journal of Sport & Social Issues*.

Docalogue

Series Editors: Jaimie Baron and Kristen Fuhs

Each book in the Docalogue book series highlights a recent documentary film from five different scholarly perspectives. By focusing on a single documentary from multiple points of view, each book demonstrates the ways in which a single film can open onto diverse questions having to do with the status of the "real," documentary ethics, and the politics of representation, among other issues. The book series is an extension of Docalogue.com, a monthly online publication that consists of short essays about contemporary documentary films.

I Am Not Your Negro
A Docalogue
Edited by Jaimie Baron and Kristen Fuhs

Kedi
A Docalogue
Edited by Jaimie Baron and Kristen Fuhs

Tiger King: Murder, Mayhem and Madness
A Docalogue
Edited by Jaimie Baron and Kristen Fuhs

Honeyland
A Docalogue
Edited by Jaimie Baron and Kristen Fuhs

For more information on the series, visit: https://www.routledge.com/Docalogue/book-series/DOCALOGUE

Honeyland

A Docalogue

Edited by
Jaimie Baron and Kristen Fuhs

NEW YORK AND LONDON

First published 2022
by Routledge
605 Third Avenue, New York, NY 10158

and by Routledge
4 Park Square, Milton Park, Abingdon, Oxon, OX14 4RN

Routledge is an imprint of the Taylor & Francis Group, an informa business

Library of Congress Cataloging-in-Publication Data
Names: Baron, Jaimie, editor. | Fuhs, Kristen, editor.
Title: Honeyland : a docalogue / edited by Jaimie Baron, Kristen Fuhs.
Description: New York : Routledge, 2022. |
Series: Docalogue 4 | Includes bibliographical references and index.
Identifiers: LCCN 2021057790 (print) | LCCN 2021057791 (ebook) | ISBN 9780367644529 (hardback) | ISBN 9780367644550 (paperback) | ISBN 9781003124573 (ebook)
Subjects: LCSH: Honeyland (Motion picture) | Documentary films.
Classification: LCC PN1997.2.H6456 H66 2022 (print) | LCC PN1997.2.H6456
(ebook) | DDC 791.43/72—dc23/eng/20211216
LC record available at https://lccn.loc.gov/2021057790
LC ebook record available at https://lccn.loc.gov/2021057791

ISBN: 978-0-367-64452-9 (hbk)
ISBN: 978-0-367-64455-0 (pbk)
ISBN: 978-1-003-12457-3 (ebk)

DOI: 10.4324/9781003124573

Typeset in Times New Roman
by codeMantra

Contents

Figures

Foreword

Docalogue began in 2017 – and continues – as an online journal, but it also began as a documentary salon in Los Angeles a decade earlier when the editors were both graduate students. Each month, we and a number of friends and colleagues would meet at one of our homes to watch and discuss a documentary film. Although the salon only lasted a year or so, it was one of the most stimulating forums for discussion of documentary film that we experienced during our graduate years. When the editors each moved on to academic jobs in different cities, we continued to meet at conferences, particularly Visible Evidence, which provides a major forum for documentary screening and discussion. Although Visible Evidence is always exciting and generative, we longed to have a way to sustain our discussions of documentary media throughout the year. From this desire arose Docalogue, a digital publication wherein we select one recent documentary each month and solicit two scholars to write a short essay about it, offering two perspectives intended to start off a broader conversation, whether on the website, in classrooms, or within documentary scholarship more broadly.

After about a year of provocative posts in this form, we decided that we might expand the Docalogue format to include short, edited books offering multiple perspectives on a single documentary film – a format that had rarely been tried, at least for nonfiction media. One of the challenges we have faced is how to decide which documentaries to choose as subjects of a book-length study. On the website, this is less pressing since we feature so many documentaries, and the purpose is simply to foster scholarly conversation. In choosing documentaries for the book series, however, we are by definition singling out particular documentaries that we think have more than passing significance. And, since our focus is on recent documentaries, this is necessarily a gamble: we do not know for certain which films will stand the test of time. In addition, while our aim is not to establish a new canon, by virtue of focusing a whole book on a film, we cannot help but raise the profile of the film at least within the documentary scholarly community. In the end, we decided to take the risk and simply choose films that we believe raise important issues about documentary in the contemporary moment and open themselves up to multiple avenues of scholarly analysis. Moreover, our aim is also to center at least some films that emerge from makers whose voices have not always been foregrounded by documentary scholarship.

The purpose of the Docalogue book series is, however, not to close the book, as it were, on any film. The idea is to open up conversation among scholars, to demonstrate to students the many ways of approaching a documentary text, and to offer a resource for those who wish to teach recent documentary films about which little has been written so far. We hope that, like the online journal, the book series will give rise to further scholarship about the films in question.

We would like to thank our Board of Advisors – Chris Cagle, Timothy Corrigan, Oliver Gaycken, Maria Pramaggiore, Pooja Rangan, Mila Turajlić, and Janet Walker – for their advice and suggestions regarding the selection of films and writers. Thanks to Natalie Foster, Sheni Kruger, Jennifer Vennall, and the whole team at Routledge for supporting this series. Our gratitude goes out to all of the writers who have contributed thus far to the Docalogue project – in both the book series and on the website.

For more information about the Docalogue, go to www.docalogue.com.

Introduction

The *I* and *Thou* of *Honeyland*

Jaimie Baron

> We live our lives inscrutably included within the streaming mutual life of the universe.[1]
>
> Martin Buber

Honeyland (Tamara Kotevska and Ljubomir Stefanov, 2019), a film about a woman of Turkish descent named Hatidže Muratova who lives with her ailing mother Nazife in a rural area of North Macedonia and makes her living harvesting honey using an ancient and sustainable form of apiculture, was a surprise hit on the film festival circuit and a significant financial success. Despite its intensely local focus, the film's enthusiastic reception in North America, Europe, and beyond suggests that its depiction of Hatidže's life spoke to many and disparate audiences living far away and in quite different circumstances.

But why? How did a film about an obscure beekeeper end up so widely seen, let alone nominated for two Academy Awards? Certainly, some of its success has to do with how documentary films circulate in the contemporary festival marketplace. Ilona Hongisto's chapter in this volume illustrates, among other things, the ways in which funding models, festival regulations, premiere locations, and distribution strategies contributed to *Honeyland*'s success. Yet, good marketing and buzz can only take a film so far. Something in the film itself must speak to audiences enough that this buzz turns into genuine popularity. Thus, while Hongisto's analysis deftly demonstrates *how* the film became so successful, an analysis of the industrial practices around the film cannot fully explain *why* it succeeded beyond expectations.

It is worth noting that *Honeyland* is a particularly difficult film to classify. Hongisto discovered that, when the film's producer inquired about including the film in the prestigious Berlinale film festival, he

DOI: 10.4324/9781003124573-1

received (and declined) an invitation to present the film in the festival's "Culinary Cinema" category. Meanwhile, when I went to look for it in my university library, the DVD was not where most documentaries are usually located but on a different floor altogether. Perplexed, I wandered through an area of the stacks I had never before visited and found the film in the "agricultural" section. These two odd placements of *Honeyland* speak to the film's categorical elusiveness. It is, by most definitions, a documentary and, within that overarching category, could be considered an ethnographic film, an observational film, or even a portrait of its main subject, Hatidže. Other ways to approach the film would be to regard it as what Chris Cagle has identified as a "popular art documentary"[2] or what Benjamin Schultz-Figueroa has called a "multi-species documentary,"[3] and these would certainly be helpful terms in unpacking the film and its significance. Yet, spurred by the Library of Congress classification system, I want to consider what it means to see it as an "agricultural film." The phrase seems to suggest a how-to of farming techniques, and – in a roundabout way – *Honeyland* does offer a basic apicultural education. However, it also proffers a different kind of education having to do with the range of possible interactions between humans, non-human animals, and other living beings, as well as between filmmaker, film subject, and film viewer.

The word "agriculture" comes from the Latin words *agrum*, meaning field or land, and *cultura*, meaning cultivation or care. While we tend to think of agriculture as synonymous with farming, this etymology points to the notion that human beings are tasked with *care* for the land – and arguably, by extension, those creatures that inhabit the land. Such care is not necessarily the same as conservation of pristine nature; agriculture – like apiculture – is about "caring for," but this does not preclude use for human ends. Indeed, while the film sets up a contrast between Hatidže's sustainable approach to beekeeping and the profit-driven apiculture practices of her neighbors the Sams, the idea that humans have a right to – indeed, must – make use of the land and its creatures is not in question here. In fact, *Honeyland* implicitly points to how a purely sentimental attitude toward non-human animals is a luxury of those – like me – who do not directly and immediately depend on them for day-to-day survival. Rather than simplistically opposing instrumentalization of nature with sentimentality, I suggest, the film constitutes an exploration of two distinct orientations between the self and other beings as articulated by Martin Buber through his immersion in Jewish mysticism: the *I-It* and the *I-Thou*.

The title of *Honeyland* evokes the Book of Exodus in which the people of Israel, wandering in the desert, are first promised they will one day arrive in the "land flowing with milk and honey." In *Honeyland*, we do see honey dripping and milk splashing, if not quite flowing. After they have arduously sawed through a fallen tree to extract an elusive hive, for instance, we see Hussein Sam and the trader Safet biting into honeycombs, the liquid dripping down their chins. In another scene, an irritated cow kicks over the bucket into which one of the Sam children has been milking her, spattering him with milk. As these scenes demonstrate, human acquisition of both honey and milk requires knowledge and work, dependent on bees and cows. Indeed, the life we see portrayed onscreen is not the one of ease implied in the Bible but one of relentless and even painful labor for both human and non-human animals. In most developed countries, the work of producing the honey we stir into our tea or the milk we pour into our coffee is invisible, obscured by global trade practices. *Honeyland* is, from one perspective then, also a labor film that exposes the means of production. Beyond this initial exposure, however, the film also reveals how differential means of production, distribution, and consumption shape the horizon of possible connections between humans and our world, particularly its other living inhabitants. While one can read the film as idealizing Hatidže and vilifying Hussein Sam, I see them, rather, as enacting two different approaches to their worlds based on the pressures under which they have been placed.

In the film, Hussein – ensnared like most of us by the forces of capitalism and harried to make money quickly – enacts an *I-It* orientation toward the bees and toward his wife and children, treating them as objects to be used. As translator Ronald Gregor Smith summarizes in his introduction to Buber's work, the *I-It* structure "presupposes a single centre of consciousness, one subject, an *I* which experiences, arranges, and appropriates. This is the characteristic world of modern activity."[4] The experiencing *I* reduces all others to *It*, to things to be used in some form by and for the self. Treated as objects, the disrupted bees sting Hussein and his children in defense of their hives. Hatidže, however, maintains a different orientation toward her world. Smith explains Buber's *I-Thou* relation thus,

> The characteristic situation here is one of meeting: I *meet* the Other. In the reality of the meeting no reduction of the *I* or of the *Thou*, to experiencing subject and experienced object, is

> possible. So long as I remain in relation with my *Thou*, I cannot experience it, but can only know it in the relation itself.[5]

One of the most striking aspects of Buber's thinking is that the *Thou* is not necessarily another human being or even an animal. It can, for instance, be a tree.[6] In the film, Hatidže embodies an *I-Thou* relation to her bees, to the other animals, to everyone she encounters. Rather than *experiencing* other beings, she *meets* all things – adults, children, animals, perhaps even trees and inanimate objects. This is not naïveté or ignorance but, rather, a refusal (conscious or no) to (fully) submit to a regime that insists that the self is the center of everything. When she speaks to the bees as "you," when she chats comfortably with the merchants in the Skopje market, when she talks to a child or to a cat, she acknowledges them as themselves.

For its part, the film not only captures this *I-Thou* relation but also solicits from the viewer something more (or other) than an *I-It* orientation toward those we see onscreen: Hatidže, Nazife, the Sams, the bees, the cows, and the land itself. Of course, an *I-Thou* relation is impossible through the mediation of film, since relation in Buber's view must be mutual, reciprocal, and in the present. When we watch a film, we are necessarily located at a spatial and temporal distance, experiencing others onscreen. As demonstrated by Linnea Hussein and Andy Rice's chapters in this volume, representing or looking at an-other onscreen is always a fraught enterprise, riddled with ethical pitfalls. Hussein argues in her chapter that, by setting up different "time zones," in which Hatidže represents an idyllic and timeless past and the Sams a recent industrial modernity (but not quite the present), the filmmakers allow viewers to feel they do not inhabit the same temporality as either. In Hussein's view, this amounts to an abdication of ethical responsibility towards both Hatidže and the Sams by both filmmakers and viewers. In a related vein, Andy Rice argues that the film can be read as an example of salvage ethnography in which the filmmakers attempt to preserve Hatidže's life and experience before they are lost forever to the forces of encroaching capitalist modernity. By treating the disappearance of her way of life as inevitable, he suggests, the film accepts a teleological view of history in which there is no place for Hatidže in the future.

These critiques are both valid and necessary. There is too much danger in accepting ethnographic representation and its consumption without rigorous and relentless self-critique on the part of filmmakers, viewers, and scholars. Yet, I want to suggest that there can sometimes – perhaps only rarely – be an asymptotic approach of the *I* to *Thou* through film. As Maja Manojlovic suggests in her chapter, *Honeyland* can also be read through the lens of what Donna Haraway calls *sympoiesis*, a process of "worlding-with, in company of" other beings, which may be another way of articulating the *I-Thou* relation. Meanwhile, Selmin Kara notes the ways in which the film demonstrates a reversibility between human and bee, offering what Kara calls an "ethological realism." By breaking down certain boundaries between the *anthropod* and *arthropod*, the film suggests a move away from an exclusively *I-It* orientation toward non-human animals, even those as seemingly alien to us as insects. According to these readings, in both its content and its formal properties, *Honeyland* offers something beyond the exhausted and exhausting ethnographic paradigm or its critique: the possibility of a categorically different way of being in the world.

Watching *Honeyland* I could not help but think of Luis Buñuel's classic but controversial 1933 film *Land without Bread.* Whereas the title of *Honeyland* suggests abundance, Buñuel's title indicates lack. While *Honeyland* suggests the promised land, *Land without Bread* (titled in one version as *Unpromised Land*) explicitly points to its opposite.[7] In both films, however, we see children suffering, women whose faces appear prematurely aged, and much death. Bees even figure briefly into Buñuel's film. According to lore, Buñuel asked one of the local people, the Hurdanos, to cover a sick donkey with honey so that bees would sting it to death. The image of the donkey covered in bees is one of many shocking images of dead animals in the film. Buñuel's film has been both derided for its negative depiction of the Hurdano people as well as celebrated as a critique of the ethnographic film at its dehumanizing worst. The fact that the film is both a dehumanizing document and a critique of such dehumanization is what makes it both disturbing and endlessly fascinating. Its meaning oscillates from moment to moment and remains, finally, undecidable. It ironically inhabits an extreme version of the *I-It* orientation toward the Hurdanos in order to make us see how we all participate in this orientation. This critique, however, does not attempt to engage the *I-Thou* relation. Such relation is indicated only by its total and conspicuous absence.

In contrast, *Honeyland* reads to me as without (intentional) irony. It appears, rather, as a sincere attempt to represent, however partially, the life-worlds of Hatidže, her mother, their neighbors, and the non-human animals with which they coexist. Whereas Buñuel's irony serves to insulate his film from certain kinds of critique, *Honeyland* does not seek this kind of distanced refuge, which leaves it quite open to attack, like a beekeeper without a mask. Yet, there is courage in this sincerity. Indeed, although I recognize the inherent problem of looking at those who are different and distant, the alternative seems to be that we disregard or remain ignorant of the lives and life-worlds of such others, which seems a poor solution. I want to argue that *Honeyland* models an unobtrusive respect towards its subjects, giving us an opportunity – albeit brief and imaginative – to *meet* them, to locate ourselves in relation to them. My own experience of the film was a sense of encountering several people in whose presence I will never stand, and of acknowledging their (ongoing and simultaneous) existence in this world we share. In *Honeyland* – both in the example of Hatidže's relations with her world and in the film form – I find a call for renewal of communion across geographical, cultural, linguistic, socioeconomic, and even species boundaries. Hatidže's joy and her struggles are not mine, but I can value the simple fact that they are as real as my own. I go to documentaries to glimpse other lives that I myself might have lived. If I had been born in a different place, at a different time, or under different circumstances, my life might have been Hatidže's and hers mine. One of the greatest potentials of art is its ability to open us to truly imagining being someone else, of living a different life. Documentary, in particular, allows us to viscerally imagine what it might be like to inhabit another body, another space, another language. There are, of course, limits to this ability to imagine being other, but the draw of *Honeyland* is not – or at least not only – objectification or exoticism. It is the reciprocity of existence.

The COVID-19 pandemic has prompted in me – and I am sure in others – a recognition of how very isolated we all are, not simply during the pandemic, though it has certainly exacerbated isolation and loneliness for many. I am referring to the basic isolation characteristic of the human condition, the fact that no matter how much we reach out to others, we are all ultimately alone. Discussing this fundamental human isolation, Irvin Yalom writes,

> There is, of course, no "solution" to isolation. It is part of existence, and we must face it and find a way to take it into ourselves. Communion with others is our major available resource to temper the dread of isolation. We are all lonely ships on a dark sea. We see the lights of other ships – ships that we cannot reach but whose presence and similar situation affords us much

> solace. We are aware of our utter loneliness and helplessness. But if we can break out of our windowless monad, we become aware of the others who face the same lonely dread. Our sense of isolation gives way to a compassion for the others, and we are no longer quite so frightened.[8]

At its best, documentary is an attempt to at least look out for the lights of one or two of those other ships. Despite all the differences between Hatidže and me, the lights of her ship and mine are interchangeable. While I will likely never meet Hatidže in the flesh – across time, space, language, culture, and a million other distinctions that remain intact – she feels to me like a friend. If this does not (cannot) constitute an *I-Thou* relation to Hatidže, it is not exclusively an *I-It* orientation either, but something in between.

Hatidže is not an ideal. Idealization still depends on an *I-It* orientation, experiencing and appropriating others rather than remaining in relation to them. Seeing Hatidže as a "noble savage" living outside of time and modernity is a form of objectification. Rather than seeing *her* as a model (and hence an idealized object), however, I see the *mutual relations* she demonstrates as the model. Her agricultural and apicultural practices – her care for land and bees – cannot be reduced simply to the buzzword of "sustainability," which is just a rebranding of capitalist exploitation. Rather, her practices of care for her world – our shared world – suggest a place from which to begin to reject the *I-It* structure upon which capitalism is founded. If we are to face the current and coming planetary changes, we must (re)awaken the *I-Thou* relation between ourselves and other humans, non-human animals, plants, and the earth itself. We are always fundamentally enmeshed in the spaces in which we live and grow, and with the plants and animals (human and not) with which we interact. Anthropogenic climate change is a testament to the fact that we have never existed outside of or distinct from the ecosystem. To care for bees and other living beings is to care for ourselves. Bill Nichols' notion of "epistephilia" has come to suggest a kind of armchair tourism, a desire to feel knowledgeable but not participate or respond.[9] While this may be an aspect of documentary consumption, I believe there is also often a genuine desire on the part of the viewer to meet, to care for an-other. If *Honeyland* can cultivate an expansion in our horizon of care and of our awareness of our inclusion within "the streaming mutual life of the universe," perhaps in this time of emphatic isolation, we may no longer feel quite so frightened.

Notes

1 Martin Buber, *I and Thou*, trans. Ronald Gregor Smith (Edinburgh: T. & T. Clark, 1937), 16.
2 Chris Cagle, "*Kedi*: Crossover Documentary as Popular Art Cinema," in *Kedi: A Docalogue*, eds. Jaimie Baron and Kristen Fuhs (New York, Routledge, 2021), 68 – 85.
3 Benjamin Schultz-Figueroa, "From Cat to Clowder: *Kedi* in the Anthropocene," in *Kedi: A Docalogue*, eds. Jaimie Baron and Kristen Fuhs (New York: Routledge, 2021), 4–18.
4 Ronald Gregor Smith, "Translator's Introduction," in Martin Buber, *I and Thou*, trans. Ronald Gregor Smith (Edinburgh: T. & T. Clark, 1937), vi.
5 Smith, "Translator's Introduction," vi.
6 Buber, *I and Thou*, 8.
7 For an exploration of the deeper significance of this alternative title, see James F. Lastra, "Why Is This Absurd Picture Here? Ethnology/Heterology/Buñuel," in *Rites of Realism: Essays on Corporeal Cinema*, ed. Ivone Margulies (Durham: Duke University Press, 2003), 185–214.
8 Irvin D. Yalom, *Existential Psychotherapy* (New York: Basic Books, 1980), 398.
9 Bill Nichols, *Representing Reality: Issues and Concepts in Documentary* (Bloomington: Indiana University Press, 1991), 178.

1 Salvaging the bees

Honeyland and the paradox of the observational fable

Andy Rice

Three weeks into production, *Honeyland* filmmakers and environmental activists Tamara Kotevska and Ljubomir Stefanov recorded a scene with beekeeper Hatidže Muratova that they knew would be central to their film. Hatidže kept a bee colony in the ruins of a house wall in the abandoned village of Bekirlija, where she lived alone with her aging mother Nazife. She was removing honeycombs in preparation to sell them to vendors at the open-air market in the city of Skopje, Macedonia about 12 miles away, which happened to be the hometown of Kotevska and Stefanov. As the filmmakers recorded a high angle, over-the-shoulder shot of Hatidže gently reaching into the hollow of the wall, she began to talk to the bees. "Half for you, half for me," she whispered. Stefanov recalled the poignancy of that moment in a question-and-answer session after a screening of the film at the Lincoln Center in New York. "Her example is actually a very rare thing," he said of Hatidže, adding that he admired that "a person that lives in such extreme circumstances, without water, without electricity, without roads and contacts with other people" could nonetheless demonstrate an important environmental message about not overusing natural resources.[1] That online promotional material for the film calls Hatidže "the last in a long line of Macedonian wild beekeepers" suggests that such an ethic is on the wane, the stuff of salvage rather than model on a capitalist planet.[2] As the film's diegesis presents Hatidže as vulnerable and alone in the ruins of a former village of diasporic Macedonian Turks, urban viewers like those hearing Stefanov's story in New York may contemplate the end of her way of life, a trope long associated with the controversial "salvage paradigm" in anthropology.

This chapter considers the re-emergence of the salvage paradigm in the form of observational films like *Honeyland*. Since the early 2000s, sensory ethnographic filmmakers working in ways much like those of Kotevska and Stefanov have reclaimed observational

DOI: 10.4324/9781003124573-2

recording techniques usually associated with 1960s documentary to represent the ambiguous textures, affects, and temporalities of the precarious industrial and pastoral life-worlds of today and screened their work in high-prestige festivals and art venues like Lincoln Center.[3] I argue that certain of these observational films have come to function culturally as fables, as simple stories that aim to inculcate their audience with a moral lesson. There is an irony to my argument. Unlike ethnographic filmmakers past and present who aim to explain cultural difference with expository voiceover layered upon the image, sensory ethnographic filmmakers strive to avoid didacticism. They may emphasize the sensory experience of texture and temporality, the "feeling of being there," the observational long take as a form of resistance to conventional media pace, or the dignity of everyday life at the margins, but they do not claim to be making fables. As a group, observational cinema folks have tended to regard "message films" as statist, conservative, reductive, propagandistic, and controlling. Observing individuals, in contrast, ostensibly allows viewers to come to their own conclusions about views seen and events felt, as individuals filmed can show range and complexity.[4] Treating the camera as an indexicality machine, argues documentary theorist Erika Balsom in her polemic for a renewed "reality-based community," observational films created by sensory ethnographic art filmmakers prioritize the "non-coded powers of lens-based capture rather than the reductive linguistic paradigm of codedness proper to theorizations of film," thereby enabling viewers of diverse viewpoints to encounter the ambiguous, profilmic world. Balsom aims to reconsider critiques of the observational mode as de facto voyeuristic, naïve, colonial, and pseudo-scientific, and to get over the fact of construction in all films.[5] But when they play as fables for viewers like those at the New York screening ruminating on the end of Hatidže's way of life, observational films do encode a moralizing message—and one that follows from the dicey impulse to salvage the exotic.

Salvage ethnographies adhere to what historian of anthropology Brian Hochman called the "fable of the mummy complex." They purport to use inscriptive media (audio recordings, photographs, and films) to preserve the language, stories, customs, cosmologies, and values of a given society or group before encroaching modernity extinguishes it or changes it beyond recognition. Arguing that preconceived racial formations determined the development of inscriptive sound and image media rather than the other way around, Hochman identified Bazin's famous argument in "The Ontology of the Photographic Image" as the crystallization of a myth central to the development of ethnographic film: that humans carry within them an instinctive drive for medial preservation uniquely satisfied by cinematic mechanical reproduction.[6] "Change mummified," in the field of ethnography, might stand as synonymous with what James Clifford critiqued as "the ethnographic present," or the disavowal of subjects' ongoing life in favor of an imagined, unchanging pre-modern existence that media can preserve.[7] The framework of always seeing particular subjects as pre-modern objectifies them, ignoring their agency to adapt, change, and survive. Some commenters on *Honeyland* have identified its salvage framework as an ethical breach along these lines. Dina Iordanova, a transnational film scholar with a specialty in Balkan film, has argued that filmmakers from relatively remote locales like North Macedonia "cannot gain attention unless they resort to self-exoticism.... If the filmmakers were to simply show the reality [of poverty], most people would not have seen this film today nor would we be talking about it here."[8] The geographical proximity of urbanite filmmakers and rural collaborator-subjects in a country like North Macedonia, in other words, does not change the basic dynamic of salvage ethnography here. Kotevska and Stefanov still aimed to record visual "exotics," as they put it, that were basically the by-products of their regional compatriots' survival struggles.[9]

Honeyland thus brings together impulses usually thought to be in tension: that of the "message film" about global environmental collapse, that of the observational documentary exploring the ambiguous textures of everyday life, and that of the salvage ethnography aiming to use media to preserve aspects of fading life practices. Reviewer Michael O'Sullivan noted that something like this tension led the film to become "a strange and curious thing: part fly-on-the-wall anthropology, part ecological fable," especially after a second family moves into the village.[10] When Hussein and Ljutvie Sam and their seven children arrive about 14 minutes into the film, they bring noise, cows, and a baseline level of chaos and hunger—grist for comparison to the peaceable Hatidže and her mother. Hussein tries to emulate Hatidže's beekeeping practice to provide for his family, but he breaks from her advice to leave half of the honey for the bees when an itinerant trader offers good money for an order of honey in bulk—200 kilos—that the family cannot sustainably produce. During a frantic stretch of honey extraction, panicked bees sting the Sam children and then devastate all other bee colonies in the village, including Hatidže's, ruining what seemed to be an amicable enough relationship between the two households. The tale seems to demonstrate a 21st-century truth about capitalism and social relations, like a fable about how greed destroys fellowship as well as ecological balance. But there is also something paradoxical to the notion of the "observational fable," which would seem to suggest that we can have non-judgmental sensory experiences and a moral, too. How and why in 2019 is this observational salvage film functioning in such a way?

The fable form in *Honeyland*: juxtaposition, excision, and structure

While historically signifying "a fictitious narrative or statement" or "a fiction invented to deceive," the Oxford English Dictionary notes that the term "fable" primarily refers to a genre of short story "devised to convey some useful lesson," often involving animals or inanimate things that speak and act, as in those of Aesop.[11] Neither meaning would seem emblematic of intent, ethic, or camera technique in observational cinema. A fable is short; durational observational filming is long. A fable aims to communicate clearly using words; observational films offer ambiguous sensations. A fable is skimpy on details and forward-projecting rather than thick and made of records of the past. Fables require conflict, while the structure of observational cinema does not, if the world viewed refuses to oblige. Yet *Honeyland*'s observational style hews to both the narrative and moral qualities of the fable.

Russian literary scholar Lidiya Vindt, writing an incisive genre analysis of the fable in the 1920s (translated into English in 1987), identified three key characteristics that cut across fables from many different times and places. First, Vindt argues, the fable is an allegorical form that signifies on a narrative plane as well as a moral plane. The genre generates expectations for readers that they will receive a lesson, a bit of wisdom applicable to life in general.[12] While it is not immediately evident that *Honeyland* will offer such a lesson, the film suggests it visually throughout and delivers that lesson by the film's end. From early on, it is clear that the lesson will be something about the destructiveness of modern ideas about efficiency. Many scenes show signifiers of modernity seeming to encroach into Hatidže's space. They especially recur in shots that provide a contemplative buffer between scenes more immediate to plot. Men wear sunglasses and look at their iPhones at a festival Hatidže and the Sams attend in the mountains outside the village. Airplanes fly overhead in the background of everyday activity in the village. The filmmakers also juxtapose natural and manufactured elements in the landscape, like a shot of vultures circling in the sky edited adjacent to a shot of one of the Sam daughters sitting on the roof of the family house with jet exhaust visible in the background (Figure 1.1). The continuous sound of rumbling jet engines across the cut suggests a

Figure 1.1 A shot of the Sam daughter sitting on top of the family house with plane exhaust in the background (top, a) precedes a shot of vultures circling in the sky (bottom, b), communicating allegorical meaning about the precarity of ecological balance.

constructed juxtaposition rather than a profilmic co-presence between the Sam daughter and the circling vultures. The editing visually communicates the fable's lesson: modernity circles like a vulture over the tenuous ecological balance for which Hatidže and her bee colonies stand. Similarly, the Sams' animal management practices contrast sharply with those of Hatidže throughout the film. Though the observational style may sometimes communicate sensory immersion and abundance, the filmmakers of *Honeyland* find ways for their imagery to function allegorically along the lines of Vindt's characterization of the fable.

Second, Vindt points out that the characters in a fable (human or animal) "must serve as symbols of human relationships." In this way, the fable plays on types and tropes which stand in for abstract values. Too much humanization and concreteness "makes [the story] ponderous, hampering its transference to other particular instances and shielding the general outline which must be apparent behind it," she notes.[13] While *Honeyland* focuses on the organic narratives that arise in the midst of filming concrete individuals, there are nonetheless telling excisions that serve to maximize the ways that these individuals can "serve as symbols." These excisions extend from the observational style and salvage orientation of the film: there is no historical context, relationship to the state, explanation of events that unfold on screen, or even a communication that the Sam family is nomadic and will return next season to the village. The structure of the film suggests that the Sams have destroyed the possibility of bee keeping here and will leave permanently. Picking up on this framing in his review, Richard Brody critiqued what he saw as the film's "hermetic schema" that focuses the film's dramaturgy for "strategic effect." "[E]liding the modernity of Hatidže—the bureaucratic ties that bind her to the administrative state, the sense of extended family, of friendships, of politics, of personal history—...establish[es] an agrarian fantasy of perfect rustic isolation."[14] The brief references to encroaching modernity like those mentioned above serve as background, not subject. The plot remains focused on the universal category of neighborly relations, demonstrated by a simplified rendition of the ongoing disputes between the Muratovas and the Sams. For all her good humor and charisma, Hatidže also symbolizes a bygone, somewhat mystical relation to local ecology here *because* the film elides her media savvy, economic thinking, political views, and limited choice in relation to her lifestyle.

Extratextual information about the making of the film and the history of Turkish villages in the region further exposes how excision bent the story of the beekeeper toward fable and away from the complexities of living with modern institutions. Hatidže has said in interviews with journalists that she identifies as an "old Turk" remaining in a diaspora in North Macedonia, though most inhabitants of ethnically Turkish, Muslim villages left after World War II. Within her family and kin group, Stefanov explained, "the last-born female child stays with the parents until their death"—a crucial detail left tacit within the film.[15] Whether intentional or not, this excision emphasizes the archetypal qualities of Hatidže's relationship to her bees over her membership within a religious and ethnic group and extended kin network. Her brothers in other villages also keep bees in skeps, but they do not appear in the film. Unlike most anthropologists, moreover, the filmmakers did not speak the language of their subjects, and so relied on image composition, body language, and tone to guide their filming choices. The complexities of talk do not lend themselves so easily to isometric representation in an image, and the filmmakers acknowledged that the translation of the footage surprised them.[16] Certainly, the afterlife of the film complicates any notion that the "hermetic schema" presented in the diegesis represented a timeless "ethnographic present." The film changed the lives of Hatidže and the Sams. After completing the project, the filmmakers used some portion of the film's profits to purchase both Hatidže and the Sam family homes in Dorfullu, a village of about 30 ethnically Turkish, Muslim families close to Bekirlija, and to set up a Honeyland foundation to connect their community to global consumers of films and honey.[17] Requests for help with social security applications, disputes over access to a well, reconnecting water for dry taps, theft of Hatidže's honey, and myriad other issues still constituted responsibilities for *Honeyland* producer Atanas Gerogiev as of August 2020. "It's impossible!" he told a reporter doing a profile on this age-old documentary problem. "We are filmmakers, not social workers." This article also revealed that disputes between the Sams and Muratovas predated filming and carried on into and beyond post-production. In fact, Stefanov and Kotevska actually recorded the neighbors in a period of "relative calm" in the longer arc of their relationship.[18] Last, Hatidže moved on from subsistence bee keeping (though she still manages some beehives) and has become something of a celebrity who now offers interviews with visiting journalists and participates in Q&A sessions at film festivals. She even met a starstruck Sarah Jessica

Parker of *Sex and the City* fame when in Hollywood for the Oscars and read the actress's fortune in coffee grounds. Such culture-crossing stories are interesting and rich in their own terms but contradictory to the film's presentation of Hatidže and her world with the Sams, the one suited to the fable form.[19]

Third, Vindt notes that the plot structure of the fable often features two characters who turn out to have antithetical goals: "the denouement destroys the line projected in the initial situation, lending it an unexpected, opposite result" that the reader might generally anticipate along the way in any case, given genre expectations.[20] The plot of *Honeyland* follows such a structure; in written form, it reads like one of Aesop's fables. A woman who tends hives always leaves half of the honey for her bees, as selling this much of the honey allows her to take care of her elderly mother without killing the colonies. Being a good neighbor, she teaches a nomadic cattle rancher and father of seven the art of apiary when his family moves into her village. But being a father of seven in need of money, he sells the honey to an itinerant trader rather than leaving half for the bees. His hungry bees kill the woman's colonies and his own before he, his family, and his cows move on to other pastures. The woman's mother dies and she is left alone in the village. Two components in antithesis, followed by an unexpected opposite result (predictable in hindsight), lead to a lesson about the state of the world: leave half, but beware of hungry neighbors, or something like this. In its juxtapositions, editorial excisions, and plot structure, then, *Honeyland* presents as a fable, though the use of observational methods may lead viewers to feel that they have found the moral rather than received it. Regardless, such feelings require an underlying cultural framework for interpretation. To complete the paradox of the observational fable, that framework must allow particular viewers—the arthouse market, in this case—to quickly interpolate critical judgments of the modern context in which they live. For better or worse, this is what observational salvage ethnography has come to offer.

The salvage paradigm and observational cinema

The salvage paradigm originated with artifact and language documentation in the 1830s when Europeans began to recognize the devastating impact of settler colonialism on indigenous peoples, but it changed significantly in spirit and orientation as anthropology professionalized and practitioners began using film cameras in their fieldwork. Whereas early ethnologists like James Cowles Prichard in the 1830s and John Lubbock in the 1860s focused on the preservation of artifacts from "savage races in other parts of the world" for a nascent comparative science of human evolution, by the 1890s Franz Boaz came to see salvage ethnography rather as crucial for helping threatened tribal members assimilate into modern society.[21] In contrast, amid pursuits of empire and global monopoly capitalism in the early 20th century, some artists, filmmakers, and anthropologists took up new recording technologies in the cause of salvage as a means to critique the societies from which they came. To such ends, Edward S. Curtis's 20-volume, decades-long documentary project *The North American Indian* (1907–1930) featured thousands of photographs, oral histories, and wax gramophone cylinder recordings of languages and songs from members of over 80 tribes.[22] In lean years, however, Curtis was not above marketing his connections to native peoples for films with sensationalistic titles like *In the Land of the Head Hunters* (1914), an unprofitable precursor to Flaherty's box office hit *Nanook of the North* (1922).[23] Flaherty, likewise, aimed to use film to visualize bygone cultural practices like harpoon hunting, and routinely employed staging and reenactment techniques to heighten the drama of his naturalist take on Inuit survival struggles—even promoting the fact of lead actor Allakarialak's death a short time after the filming as a marker of the film's authenticity.[24] Filmed and distributed prior to any formalization of the term "documentary" and associations with filming everyday life in its unfolding, distributors marketed *Nanook* as an independent film by a sensitive artist, at odds with the crass melodrama of studio fiction. In keeping with nascent ideas about proper ethnographic fieldwork, Flaherty presented himself as a collaborator who cultivated longstanding relationships with subjects, albeit in the service of scripted drama meant to evoke the sense of life as it, in his view, once must have been.[25]

After the formation of anthropology programs in many Western universities, ethnographic researchers incorporated film cameras into their research, and in doing so began to envision salvaging different kinds of traces. Across her 50 years of research from the 1920s to the 1970s, Margaret Mead came to see cameras as vital for creating "record footage" of interpersonal interactions otherwise lost to flux, the kind of everyday life details difficult to detect in person and describe in fieldnotes. Recording "disappearing types of behavior" was an urgent activity among family groups in cultures undergoing rapid change, she thought, both to "permit the descendants to repossess their cultural heritage" and to "give our understanding of human history and human potentialities a reliable, reproducible, reanalyzable corpus."[26] But Mead was no observational filmmaker. She insisted that only expert commentary over images could make them legible for colleagues and the broader public unfamiliar with subject groups, and therefore usable as tools for education about cultural difference.

By the 1970s, when direct cinema and cinema verité had gained traction as dominant realist cinema movements and Bazin's writings on realism had been translated into English, observational cinema proponents like Colin Young argued that the image alone, skillfully recorded by trained camerapersons, informed by academic discipline and extended research, and edited so as to follow everyday life events in subjects' worlds, could do a better kind of educational work than expository forms.[27] The observational form of film could teach viewers how to think for themselves about images, sounds, sensations, and individuals on screen—habits of thought imagined by thinkers like Young in that fragile post-World War II period to be vital to the resilience of a democratic society—and lead viewers to a kind of "transcultural" empathy, as Young's influential protégé David MacDougall would come to call it.[28] What they "salvaged" was thus somewhat more difficult to categorize than the records of Curtis or Mead, something more broadly about, as MacDougall put it, "what we have most valued" as visual anthropologists who had experienced "close personal acquaintance with a particular society." These were "persons whom we met, certain rooms and streets and compounds where they lived, journeys taken, dilemmas addressed, objects made and used, sounds heard, faces and conversations, fears and pleasures"—subjective sensations remembered vividly.[29] These were the unexpected details of encounter that some critical cultural theorists of the early 2000s began to describe as *affective* rather than simply informative.

However, for authors of 1980s and 1990s cultural studies analyses of race, gender, and class injustices suffered at the hands of white men situated in powerful modern institutions, the prospect of anthropologists scrutinizing the filmed bodies of the world's poor for ill-defined revelations about who "we" are registered as exploitative rather than inspiring. This writing precipitated the death knell for salvage projects.[30] For filmmaker and cultural theorist Fatimah Tobing Rony, Flaherty's work—and that of other salvagers past and present by extension—was a form of unethical taxidermy, of making "the dead look alive and the living look dead."[31] Celebrations of salvage films by Western reviewers at the time of *Nanook*'s release, Tobing Rony pointed out, reflected more about the erotic desires of the West for a fascinating Other than particular concerns for improving the lives of the peoples' portrayed. Theorist Michael Renov applauded, meanwhile, the emergence of new autobiographical documentary forms across the 1980s and 1990s that undermined the way observational cinema "mimes the omniscience of conqueror."[32] Filmmaker Jill Godmilow excoriated the fetishization of the profilmic real as "documentary's albatross" in her 2002 manifesto "Kill the Documentary as We Know It." Her parodic bullet point manifesto of "do nots" for her fellow documentary filmmakers specifically targeted films that "reproduce the surface of things [without making] an intelligent proposition," "celebrate the 'old ways' and mourn their loss," or "produce freak shows of the oppressed, the criminal, the different, the primitive."[33]

There is a strong sense in such critiques that the premises of observational filming and the salvage paradigm were fundamentally flawed—even irredeemable. "When you work with documentary images," Godmilow concluded, "unless you purposefully contradict them within the text of your film, the claim will always be made that these images, taken from life, accurately represent the real world." An ideologically coherent, critical film must undermine the very presentation of the image as representative. Observational films remain committed to valuing the presentation as such of things seen, persons met, and journeys taken, to paraphrase MacDougall. Unlike the visual anthropologists seeking redeeming value in Flaherty, Godmilow asked, rather, "How do we avoid making *Nanook of the North* every time we pick up the camera—that is the question for me."[34]

Yet around the same time that Godmilow posed this question, curiously, the field of anthropology again began to consider observational film as a compelling research form. In part, the turn to affect led by critical scholars like Brian Massumi and Eve Sedgwick in the early 2000s happened to dovetail with the longstanding interests of observational filmmaker-anthropologists.[35] Practitioner-theorist and then editor of the *Visual Anthropology Review* Lucien Castaing-Taylor and his partner Ilisa Barbash, aiming to connect the dots, proposed that the inherent excess of the observational image beyond coded language could offer up the grounds for a "postsemiotic anthropology" centered on the ethnographic study of embodiment, sensory experience, and affect. They proposed through their work for the Sensory Ethnography Lab (SEL) at Harvard that observational filming of societies, individuals, neighborhoods, and work sectors undergoing transformative change could be a better means of creating and sharing such knowledge than writing theory.[36] The couple even reclaimed the "totally retrograde" term "salvage ethnography" to describe their elegiac SEL foray *Sweetgrass*, which follows Montana cowboys on one of their last sheepherding drives into the public grazing lands of the Absaroka Beartooth mountains. "It was a chance, or a challenge, for us to engage anew with 'salvage ethnography'—how to represent a world on the wane," they told Scott MacDonald in a 2012 interview about the film. "Could we acknowledge a historical loss, without falling prey to all the pitfalls of patronizing romanticism and nostalgia?"[37] Thematically, this film ruminates on the decline of the cowboy (and a particular style of American masculinity more broadly), long the emblem of rugged American individualism for some and scorched earth land management for others. If some viewers expressed nostalgia for a life on the range they never personally knew (like those viewers of *Nanook* before them), they also noted the peculiar way that the "slow cinema," observational style of the film seemed to evoke foreign temporalities and life-worlds.[38] Proponents in anthropology and beyond suggested that perhaps these unfamiliar "structures of feeling," to use Raymond Williams' resonant phrase, could lead to novel forms of political subjectivity, like a cinematic equivalent to the "slow food" movement emerging to "rescue extended temporal structures from the accelerated tempo of late capitalism."[39]

The object of salvage, in other words, was no longer culture, gesture, custom, or concept of something "pre-modern" exactly; it was the phenomenon of slowness itself imagined as an approximation of a more meaningful real than the discourses on offer in a fast-moving, slipshod social media ecosystem. If slowness were the subject, proponents argued circa 2009, then observational film was not about omniscience and conquest or neutral, scientific data gathering, but about "attending to the world—actively, passionately, concretely—while, at the same time, relinquishing the desire to control, circumscribe or appropriate it."[40] Such films re-emerged as film festival darlings in the 2010s, emblems of resistance highly valued within what SEL alumna turned critic Toby Lee ambivalently identified as a "prestige economy."[41] For Balsom, the kind of documentary that Godmilow and Renov celebrated for deconstructing its own imagery in essayistic voiceover "contains a whiff of cynicism" in a post-truth discursive context, at least compared to new observational cinema including the SEL's *Leviathan* (2012),

Manakamana (2013), and *The People's Park* (2012), installations by Harun Farocki, and Kevin Jerome Everson's unconventional *Tonsler Park* (2017).[42] In proposing this "reality-based community," Balsom admits being engaged in a kind of rescue operation. "I, too, attended all those graduate school seminars in which we learned to deconstruct Enlightenment principles and mistrust empiricism," she acknowledges, "but given the state of things, it's starting to look like they might need salvaging."[43]

Conclusion

So what are we to make of the contradiction of the observational film, celebrated for "creating for the viewer a time and space of attunement in which a durational encounter with alterity and contingency can occur, with no secure meaning assured," functioning in 2019 to reanimate the salvage paradigm and express a fable?[44] We might account for this convergence by understanding slowness and sensory experience of the everyday to be the fulcrum of an oppositional media politics and observational films rooted in extended periods of ethnographic fieldwork like *Honeyland* to be central mediating objects in its production and proliferation. We might soak in this "structure of feeling" in moments when the deer unexpectedly dashes out of a bush as Hatidže walks by, the grain of Nazife's voice telling her daughter flatly that she has become a tree, or the negotiation of injury among the Sam children and their mother when one pushes a second onto a stump. Such moments of contingency defy *a priori* knowledge or expectation beyond the improvisational craft of camerapersons attuned to the rhythms of people they know intimately, if differently, from co-present interactants not carrying cameras. It is possible to regard such moments as a viewer with something other than empty humanism or voyeuristic fascination. This is not the "faithful fortress of experimental cinema" producing alienation effects and the heady sensation of displeasure for ever-smaller audiences, but neither is such viewing simply retreading the tropes of for-profit, industrially-produced fiction.[45] *Honeyland* does impose a simplistic fable onto a complex entanglement of lives and reproduces certain discourses of the salvage paradigm (especially in its marketing), but it is also a hard-won, rare, and resonant film. Indeed, *Honeyland* plays to me as a sincere effort to ward off cynicism and nihilism about the environment and a relation to the land. In many ways, it serves as an homage to its subjects and a call to the wider global community to learn something from them.

Nevertheless, we must also ask a question about the boundaries of the community proposed and the political utility of the fable form. Observational cinema techniques may be having a moment, reflexive filmmakers may deploy them on occasion in their daily practice, and they may not be as bad an object as claimed by prior eras of documentary theorists, but they are inadequate tools for ethically engaging with many vital aspects of 21st-century life, and there are thus limitations to theorizing observational practice as the basis for community, embodied learning, institution building, and politics.[46] I agree with Godmilow that filmmakers should still seek to avoid making *Nanook of the North* when they pick up their cameras, so it is alarming to me that Iordanova frames *Honeyland* as "*Nanook* for the 21st century" for the way the film represents "underdevelopment as an enchanting morality tale."[47]

There may be a silver lining on offer in *Honeyland* that was not in *Nanook*, however. By taking on the fable form, the film implicates its viewers in a different way. The Sams serve as Vindt's "antithesis," but they also stand for us who watch the film when we understand their economic calculus. The film takes us a bit by surprise, as does life, when the Sams show up, and do all manner of things that make sense to those of us who identify with their decision making. They are our surrogates, even if (or especially when) we judge them negatively. In this way, *Honeyland* opens a backdoor space for reflection on the faults in our own relation to ecology in this calamitous time. When we long to live up to Hatidže's standards as presented in the film, this sense of self-reflection within the fable may move us to take some small lesson for our own human colonies, where we've failed to leave half.

Notes

1 Film at Lincoln Center, *Tamara Kotevska and Ljubomir Stefanov on Honeyland, Family & Capturing Macedonia | NDNF19*, 2019, https://www.youtube.com/watch?v=_ES4mfYYsII.
2 "Story – HONEYLAND," accessed May 19, 2021, https://honeyland.earth/story/.
3 Linnea Hussein's contribution to this volume offers a genealogy of and rationale for the observational style in sensory ethnography, so I am minimizing this background here. There is an excellent recent summary and critique of the sensory ethnographic film movement in Toby Lee, "Beyond the Ethico-Aesthetic: Toward a Re-Valuation of the Sensory Ethnography Lab," *Visual Anthropology Review* 35, no. 2 (2019): 138–47, https://doi.org/10.1111/var.12184.
4 See key arguments for this school of filmmaking in David MacDougall, *The Corporeal Image: Film, Ethnography, and the Senses*

(Princeton, NJ: Princeton University Press, 2006); David MacDougall and Lucien Castaing-Taylor, *Transcultural Cinema* (Princeton, NJ: Princeton University Press, 1998); Anna Grimshaw and Amanda Ravetz, *Observational Cinema: Anthropology, Film, and the Exploration of Social Life* (Bloomington; Chesham: Indiana University Press : Combined Academic [distributor], 2010); Lisa Stevenson and Eduardo Kohn, "'Leviathan': An Ethnographic Dream," *Visual Anthropology Review* 31, no. 1 (2015): 49–53; Scott MacDonald, *American Ethnographic Film and Personal Documentary: The Cambridge Turn* (Berkeley: University of California Press, 2013); Colin Young, "Observational Cinema," in *Principles of Visual Anthropology*, ed. Paul Hockings, Third (Boston, MA: Walter de Gruyter, 2003), 99–114; David MacDougall, "Beyond Observational Cinema," in *Principles of Visual Anthropology*, ed. Paul Hockings, Third (Boston, MA: Walter de Gruyter, 2003), 115–32; David MacDougall, "Observational and Participatory Approaches. Colin Young, Ethnographic Film and the Film Culture of the 1960s," in *Memories of the Origins of Ethnographic Film*, ed. Beate Engelbrecht (Frankfurt am Main; New York: Peter Lang, 2007); Ilisa Barbash and Lucien Castaing-Taylor, *The Cinema of Robert Gardner* (Oxford; New York: Berg, 2007); Ricky Leacock, "A Search for the Feeling of Being There," May 20, 1997, https://mf.media.mit.edu/courses/2006/mas845/readings/files/RLFeeling%20of%20Being%20There.pdf.

5 Erika Balsom, "The Reality-Based Community," *E-Flux Journal* 83 (June 2017), https://www.e-flux.com/journal/83/142332/the-reality-based-community/.

6 Brian Hochman, *Savage Preservation: The Ethnographic Origins of Modern Media Technology* (Minneapolis: University of Minnesota Press, 2014), xv–xvi.

7 James Clifford, *The Predicament of Culture: Twentieth-Century Ethnography, Literature, and Art* (Cambridge, MA: Harvard University Press, 1988).

8 Dina Iordanova, "Underdevelopment, Coated in Honey," *DOCALOGUE*, June 1, 2020, https://docalogue.com/honeyland/.

9 Film at Lincoln Center, *Tamara Kotevska and Ljubomir Stefanov on Honeyland, Family & Capturing Macedonia | NDNF19.*

10 Michael O'Sullivan, "'Honeyland' Review: A Documentary about a Macedonian Beekeeper Delivers a Message about Life out of Balance," *The Washington Post*, August 5, 2019, https://www.washingtonpost.com/goingoutguide/movies/in-honeyland-a-solitary-beekeeper-delivers-a-warning-about-life-in-and-out-of-balance/2019/08/05/838f40f8-b491-11e9–8949-5f36ff92706e_story.html.

11 "Fable, n.," in *Oxford English Dictionary Online* (Oxford University Press), accessed May 19, 2021, http://www.oed.com/view/Entry/67384.

12 Lidiya Vindt, "The Fable as Literary Genre," trans. Miriam Gelfand and Ray Parrott, *Ulbandus Review* 5 (1987): 88–108.

13 Vindt, 90, 98.

14 Richard Brody, "*Honeyland* Reviewed," *The New Yorker*, August 1, 2019, https://www.newyorker.com/culture/the-front-row/honeyland-reviewed-a-gripping-frustrating-documentary-about-a-beekeepers-fragile-isolation

15 Tamara Kovacevic, "Honeyland: Life Lessons from Europe's Last Wild Beekeeper," *BBC News*, February 9, 2020, sec. Entertainment & Arts, https://www.bbc.com/news/entertainment-arts-51401315.
16 Film at Lincoln Center, *Tamara Kotevska and Ljubomir Stefanov on Honeyland, Family & Capturing Macedonia | NDNF19.*
17 Mustafa Ozturk, "Documentary Carries Macedonian Beekeeper out of Poverty," *Anadolu Agency*, January 23, 2020, https://www.aa.com.tr/en/culture/documentary-carries-macedonian-beekeeper-out-of-poverty/1711669.
18 Patrick Kingsley, "Film Crew Spent 3 Years in Remote Balkan Hamlet. Will They Ever Leave?," *The New York Times*, August 29, 2020, sec. World, https://www.nytimes.com/2020/08/29/world/europe/honeyland-north-macedonia-bees.html.
19 "Hatidže, the Macedonian Beekeeper Charming Hollywood," France 24, February 6, 2020, https://www.france24.com/en/20200206-Hatidže-the-macedonian-beekeeper-charming-hollywood. d
20 Vindt, "The Fable as Literary Genre," 90.
21 Jacob Gruber, "Ethnographic Salvage and the Shaping of Anthropology," *American Anthropologist* 72, no. 6 (1970): 1293–94, 1297.
22 Kalina Kukiełko-Rogozińska, "Following the Footprints of Edward S. Curtis: A Tale of the Vanishing Race," *Przegląd Socjologii Jakościowej* 16, no. 2 (May 31, 2020): 38–39, https://doi.org/10.18778/1733-8069.16.2.03.
23 Hochman, *Savage Preservation*, 118.
24 See especially Claude Massot and Sebastien Regnier, *Nanook Revisited* [Videorecording], (Princeton, NJ: Films for the Humanities and Sciences, 2005).
25 Jay Ruby, "Introduction: A Reevaluation of Robert J. Flaherty, Photographer and Filmmaker," *Studies in Visual Communication* 6, no. 2 (June 1980): 2–4, https://doi.org/10.1111/j.2326-8492.1980.tb00036.x; Hochman, *Savage Preservation*, 117–19; MacDougall, *The Corporeal Image*, 250–51.
26 Margaret Mead, "Visual Anthropology in a Discipline of Words," in *Principles of Visual Anthropology*, ed. Paul Hockings, Third (Boston, MA: Walter de Gruyter, 2003), 8–9.
27 Young, "Observational Cinema"; MacDougall, "Observational and Participatory Approaches. Colin Young, Ethnographic Film and the Film Culture of the 1960s."
28 David and Judith MacDougall graduated from Young's short-lived MA program in Ethnographic Film at UCLA in the late 1960s and became leading makers and theorists in the decades to come. See especially MacDougall, "Observational and Participatory Approaches. Colin Young, Ethnographic Film and the Film Culture of the 1960s." For ideas about the role of post-World War II indexical art in fostering democratic habits of mind, see Louis Menand, *The Free World: Art and Thought in the Cold War* (New York: Farrar, Straus and Giroux, 2021); Fred Turner, *From Counterculture to Cyberculture: Stewart Brand, the Whole Earth Network, and the Rise of Digital Utopianism* (Chicago, IL: University of Chicago Press, 2006).
29 MacDougall, *The Corporeal Image*, 273.
30 See especially from this era Michael Renov, *The Subject of Documentary* (Minneapolis: University of Minnesota Press, 2004); Michael Renov, ed.,

Theorizing Documentary (New York; London: Routledge, 1993); Alexandra Juhasz and Jesse Lerner, *F Is for Phony: Fake Documentary and Truth's Undoing* (Minneapolis: University of Minnesota Press, 2006); Laura U. Marks, *The Skin of the Film: Intercultural Cinema, Embodiment, and the Senses* (Durham, NC: Duke University Press, 2000); Faye D. Ginsburg, Lila Abu-Lughod, and Brian Larkin, *Media Worlds: Anthropology on New Terrain* (Berkeley: University of California Press, 2002).

31 Fatimah Tobing Rony, *The Third Eye: Race, Cinema, and Ethnographic Spectacle* (Durham, NC; London: Duke University Press, 2004), 126.

32 Renov, *The Subject of Documentary*, 142; MacDougall, *The Corporeal Image*.

33 Jill Godmilow, "Kill the Documentary as We Know It," *Journal of Film and Video* 54, no. 2/3 (2002): 4–5.

34 Godmilow, 7.

35 See a concise summary of this turn in Brian L. Ott, "Affect in Critical Studies," *Oxford Research Encyclopedias*, 2017, https://doi.org/10.1093/acrefore/9780190228613.013.56.

36 Barbash and Castaing-Taylor, *The Cinema of Robert Gardner*, 9.

37 Scott MacDonald, "Ruminating on Sweetgrass: An Interview with Ilisa Barbash and Lucien Castaing-Taylor," *Framework: The Journal of Cinema and Media*, 2012, https://www.frameworknow.com/news/interview-with-ilisa-barbash-and-lucien-castaing-taylo.

38 Cinema Scope editor Jay Kuehner, for instance, described the film having "a sense of purpose that may seem anachronistic, even arch, by the slack standards that constitute a current film climate which privileges the sensational over the durational." Jay Kuehner, "Interviews | Keeper of Sheep Lucien Castaing-Taylor on Sweetgrass," *Cinema Scope*, accessed May 20, 2021, https://cinema-scope.com/cinema-scope-magazine/1107/.

39 Stevenson and Kohn, "Leviathan," 49; Raymond Williams, *The Long Revolution* (Orchard Park, NY: Broadview Press, 2001), 319; Tiago De Luca and Nuno Barradas Jorge, "From Slow Cinema to Slow Cinemas," in *Slow Cinema*, ed. by Tiago De Luca and Nuno Barradas Jorge (Edinburgh: Edinburgh University Press, 2016), 3.

40 Grimshaw and Ravetz, *Observational Cinema*, 4, 5, 8.

41 Lee, "Beyond the Ethico-Aesthetic."

42 *Labour in a Single Shot: Harun Farocki/Antje Ehmann*, 2017, http://publikationen.ub.uni-frankfurt.de/frontdoor/index/index/docId/43833; Lucien Castaing-Taylor et al., *Leviathan*, 2012, https://torontopl.kanopy.com/node/6554379; Stephanie Spray and Pacho Velez, *Manakamana* (London: Dogwoof Digital, 2013); Libbie Dina Cohn and J.P. Sniadecki, *People's Park* (United States: Cinema Guild, 2012); Kevin Jerome Everson, *Tonsler Park: A Film*, included in *Second Run - How You Live Your Story*, DVD (London: Second Run DVD, 2017), https://www.secondrundvd.com/release_kevin.html.

43 Balsom, "The Reality-Based Community."

44 Balsom.

45 Jacques Rancière, *Film Fables* (London: Bloomsbury Academic, 2016), 3.

46 Observational cinema techniques offer limited tools for representing, for instance, traumatic experience, history, memory, solidarity, private

life, simulation events, and computer-based work. A full exploration of these limitations is beyond the scope of this chapter, but key arguments about such challenges may be found in: Janet Walker, *Trauma Cinema: Documenting Incest and the Holocaust* (Berkeley: University of California Press, 2005); MacDougall and Castaing-Taylor, *Transcultural Cinema*; Renov, *The Subject of Documentary*; Steven Shaviro, "Slow Cinema Vs Fast Films," *The Pinocchio Theory* (blog), May 12, 2010, http://www.shaviro.com/Blog/?p=891; Jane Gaines and Michael Renov, eds., *Collecting Visible Evidence* (Minneapolis: University of Minnesota Press, 1999); Andy Rice, "Being There Again: Reenacting Camerawork in 'In Country' (2014)," *JumpCut: A Review of Contemporary Media* 58 (Spring 2018), https://www.ejumpcut.org/archive/jc58.2018/RiceVietnamEnactment/index.html. For a critical consideration of the ways that new sensory ethnographic films neglect the co-construction of "ethical/political relationships" with film subjects, see Faye Ginsburg, "Decolonizing Documentary On-Screen and Off: Sensory Ethnography and the Aesthetics of Accountability," *Film Quarterly* 72, no. 1 (2018): 42.

47 Iordanova, "Underdevelopment, Coated in Honey."

2 Ethological realism in *Honeyland*

Selmin Kara

Tamara Kotevska and Ljubomir Stefanov's debut feature *Honeyland* (2019) opens with a drone shot of a middle-aged woman's ascent to a cliffside beehive, where she extracts honeycombs from within a buzzing crevice in the rock wall that is populated by a robust colony of European wild honey bees. She is Hatidže Muratova, a beekeeper of Ottoman-Turkish descent, living an ascetic life with her ailing mother Nazife in the mostly abandoned Bekirlija village of North Macedonia. From the beginning, the juxtaposition between the aerial shots of the barren mountain landscape, the solitary human figure who hikes across its arid expanse, and the vibrant commotion that suddenly emerges from behind a slab of rock that Hatidže gently removes upon reaching the hive presents the viewer with an ecology that demands multi-sensory attention. As the colors and patterns of the beekeeper's ocher shirt, flowery green scarf, and brown skirt blend into the golden-yellow chromatic palette of the honeycombs and thronging bees, the multi-sensory aesthetic (combining vivid imagery, sounds of the hive, and the proprioceptive sensation of flying invoked by the drone footage) appears to find a focus. That is, it pinpoints a kind of ingenious biomimicry that blurs the boundary between human and bee, bee and biome, singularity and multitude.

This chapter takes such boundary-blurring between the lifeworlds of wild bees and the documentary's human subjects, introduced in the opening sequence and sustained throughout the film, as the instantiation of what I call "ethological realism." In their discussion of a crisis of representation that has taken hold of the sciences and the arts in recent decades – under the shadow of accelerated environmental degradation – Lynn Badia, Marija Cetinić, and Jeff Diamanti propose considering different "aesthetic modes of detailing" the world that pay attention to the distributed complexity of climate change as challenging and expanding realism's

DOI: 10.4324/9781003124573-3

limits in the 21st century. Among prominent climate-aware aesthetics that have emerged, they mention "speculative, haptic, capitalist, indigenous, petro-magical, perspectival, apocalyptic, and scientific realisms" in addition to other "climate realisms" offered by schools of thought such as object-oriented ontology, new materialism, materialist feminism, and critical realism.[1] In what follows, I argue that the blurred boundary between human and bee worlds in *Honeyland* posits a mode of detailing the world that constitutes a distinct realism along the same lines. What this distinct approach offers is an ethological posturing that, when it becomes the qualifier of the film's realism, explores the relationality between the traits, behaviors, and habitats of humans and animals as something that has the potential to unsettle and shift extractivist attitudes toward the Earth's dwindling resources.[2] In doing so, it also negotiates growing concerns around biodiversity loss. Since ethological realism decenters human perspectives in favor of revealing the connections between different species' life-worlds, it suggests biodiversity loss to be the precursor to our own extinction, which the film deems avoidable if we adopt more equitable practices.

A tale of two life-worlds

By sustaining a human-bee parity throughout the narrative, *Honeyland* urges viewers to reconsider their knowledge of the world, undoing hierarchical distinctions between human and non-human subjects. While physiologically disparate, humans and bees exist in a symbiotic relationship that is mostly invisible. The European honey bee, *Honeyland*'s animal subject of interest, alone "provides nearly half of all crop pollination services worldwide" and plays a key role in eco-systemic stabilization, in addition to producing goods that are widely consumed, such as honey, beeswax, pollen, and royal jelly.[3] These ecological and material benefits give the honey bee a special status among arthropods, which often occupy a place of otherness in the popular imagination. Charlotte Sleigh explains that the apprehension humans tend to feel toward arthropods, the family of invertebrate animals named after their jointed and segmented bodies, lies in the perceived alterity of their physiology. "Insects are all wrong," she states. "There is a good case for regarding them as zoology's Other, the definitive organisms of différance. We humans have skeletons; they keep their hard parts [exoskeleton] on the outside, and the squishy bits in the middle."[4] Despite sharing such "inside out" physiology, bees have historically been viewed through a relatively more positive lens than other arthropods, as valued members of the "ecological guild" of humanity, a system of mutual support built through apiculture that has allowed the anthropos and arthropod worlds to co-exist peacefully across millennia.

In the past 15 years, however, the symbiotic relationship between humans and bees has suffered major blows. Honey bee colony deaths have become an alarming trend, with the species facing "mortality rates of 40 percent or more" each year.[5] That makes the European honey bee one of the central figures in discussions of both human food security and the mass species die-offs in the 21st century, which are concentrated among invertebrate organisms.[6] *Honeyland*'s origin story is rooted in these discussions, too. The film reportedly "began as a short [environmental] video commission from Macedonia's Nature Conservation Programme [NCP], which wanted to explore issues of biodiversity and sustainability."[7] A concern for biodiversity loss is also the central focus of the NCP's funding body, the Swiss Agency for Development and Cooperation (SDC), one of the main objectives of which is "to maintain agrobiodiversity, drawing on the knowledge and traditions of local populations."[8] The SDC acknowledges that their work is informed by the UN Convention on Biological Diversity (CBD), also known as the Nagoya Protocol, calling for "the conservation of biological diversity, the sustainable use of its components and the fair and equitable sharing of the benefits arising out of the utilization of genetic resources."[9] These are also the principles that *Honeyland* promotes, depicting them as takeaways from Hatidže's traditional lifestyle, values, and conflicts, rather than as themes that the filmmakers set out to make a film about.

Addressing the CBD directly, Stefanov describes his interactions with Hatidže – who the directors met during field research and decided to make the center of a longer, character-driven film – as reminding the crew of the principles of the Nagoya Protocol. Suggesting that the UN convention-driven logic was behind the documentary's inception, he explains:

> The scene where she takes the honey and says to the bees, "Half for me, half for you"—we shot that during the first week of filming and it was clear for us that it would be the main environmental point in the film. That "half for me, half for you" is basically the Nagoya Protocol and one of the United Nations Millenium goals.[10]

In fact, the press kit of the film places information regarding the Nagoya Protocol centerstage, by including it in the directors' statement:

> The Nagoya Protocol – a United Nations Convention on Biological Diversity (CBD) – came into force at the end of 1993 and established global guidelines on access to natural resources. Its objective was the promotion of fair and equitable sharing of benefits for both providers – i.e. land, plants, animals – and users – i.e. humans – of resources. Genetic diversity, or biodiversity, enables populations to adapt to changing environments and a changing climate, contributing to the conservation and sustainability of resources. The "honey crisis" in this film illustrates the risk of ignoring these protocols and upsetting the respect for biodiversity.

This statement, which takes "protocols" as signifying both the global guidelines on resource sharing and nature's own regulatory arrangements such as genetic diversity, informs the film and shapes its imaginary. However, instead of expository storytelling that foregrounds scientific and educational language (or what Bill Nichols calls "discourses of sobriety"), *Honeyland* adopts a lyrical, fly-on-the-wall approach to depict its vision.

What this approach – the nonintrusive camerawork, natural lighting, and no voiceover narration or talking heads formula of observational films – offers, on the surface, is a "paradise lost" kind of fable about a mother and apiarist daughter living a simple life in peaceful countryside until the creeping threats of modernity and human greed force their way into the two women's harmonious world. The conflict that arises with the arrival of a nomadic cattle-herder family, the Sams, who cause a breach in the natural order (and neighborly contract) by trying to make quick money from harvesting honey unsustainably, gives the story an archetypal quality. Notably, a similar archetypal conflict drives the plot of Norman McLaren's Academy Award-winning short documentary *Neighbours* (1952) in which neighborly conflict serves as a parable for the consequences of greed. Beneath the surface of this somewhat familiar, humanist narrative and cautionary tale, however, *Honeyland* persistently draws parallels between the lives of its human and animal subjects, making the film the meeting ground of anthropological and arthropodological life-worlds.

Here, what I mean by life-world is something akin to biologist Jakob von Uexküll's formulation of "umwelt": the world as it perceived and acted upon by different human and non-human agencies.[11] In the foreword to his influential book, *A Foray into the Worlds of Animals and Humans*, Uexküll proposes thinking of the word "environment" in the plural and beyond a strictly human perspective. More specifically, he argues that each species has a monistic or self-centered world, in which its organism-specific sensory organs present a distinct perceptual universe and its motor organs enable a particular set of actions. Every time one species interacts with another species, it steps into other unique life-worlds, in which its previous surroundings are perceptually reconfigured. When imagined as construed by bodily perceptions, actions, and relations of different species, the environment as a concept loses its singularity and conjures up an image of the world as consisting of overlapping bubble-like environments interacting and impacting one another. By rejecting the idea of a human-centered world, this reconceptualization also collapses the dualistic distinction between humans and animals, paving the way for an ontology that steers toward ethology: a thinking of reality and being-in-the-world through animal behavior.

In his study of Uexküll's biology work, which later became a pioneering framework for animal studies, Brett Buchanan attributes the popularization of ethology as a concept to zoologist Konrad Lorenz but suggests the Uexküll-influenced line of philosophical inquiry adopted by the formative thinkers of the 20th century – including Martin Heidegger, Maurice Merleau-Ponty, Gilles Deleuze, and Giorgio Agamben – to be onto-ethological.[12] What he means by onto-ethology is a perspectival shift that occurred in philosophy after Uexküll, crystallized perhaps most memorably by Deleuze and Guattari's concept of "becoming animal," an understanding of "life that is no longer based on human subjectivity" yet is valid for human beings.[13] Film scholar Elena del Rio finds ethological thought applicable to cinema as well, as an interpretive framework that foregrounds human-animal-habitat relationships and environmental ethics. Ethology, she states, "may be thought of as the examination of the affects and capacities that bind us to our habitat through a multiplicity of relations with the affects and capacities of other bodies."[14] Certainly, the life-worlds of honey bees and the two beekeeping families in *Honeyland* are interwoven and reconfigured in ways that are evocative of these ethological perspectives. Yet, I want to go further and frame the film's approach as constituting ethological realism: an approach that resonates not only with the aforementioned perspectives but also with 21st-century documentary's rekindled interest in science-reality-aesthetic relations under the ecological crisis.[15] *Honeyland* foregrounds the shared inhabitation of a destabilized environment, rather than merely documenting humans who happen to live and work with bees. The film's heightened attention to habitat disruption is informed by the findings of sciences (such as ecology, zoology, biogeography, and genetics) that track biodiversity loss in the contemporary era. The title of the film highlights this attention too. The humans and bees become, to a degree, equivalent "figures" against the dominant "ground" of the honeyland. The human-bee parity in such context offers more than mere analogy. It speaks to the ways in which the surmounting crisis of resource share among species, which threatens the well-being of all, is made legible through a documentary realism that decenters the human, as part of its ethics of nature.

The human-bee parity can be traced in the film most overtly through *Honeyland*'s common association of the mother-daughter relationship with a kind of hive-minded, dedicated care. The press kit indicates:

> Just as worker bees spend their entire lives taking care of the queen bee which never leaves the hive, Hatidže has committed her own life to the care of her blind and paralyzed mother, unable to leave their ramshackle hut.[16]

The analogy invoked in this statement is repeated by the documentary's creative team in various interviews. For example, one of the film's two cinematographers, Samir Ljuma, mentions how their avoidance of artificial lighting in the scenes that take place in the Muratova family's dimly illuminated hut allowed them to show the similarities between Hatidže's dutiful life and that of the laborious bees in their dark caverns. On the one hand, he admits that the choice of natural light was born out of necessity, as the lack of electricity in the village made charging the equipment batteries difficult, limited the crew's aesthetic choices, and called for creative solutions.[17] On the other hand, what the crew learned from such conditions of scarcity was a lesson that resonates with Hatidže's "take half leave half" mantra: "we don't need much to exist and to be happy." Yet, while Ljuma's take valorizes ideas like frugality and resourcefulness, Hatidže's hive-minded care for her mother gives rise to multiple interpretations. Before delving into how such formulations of human-bee parity might also point to an ethological realism, it is helpful to consider the ways in which various anthropocentric or anthropomorphizing interpretive frameworks have been applied to the film.

Epistemologies of the hive

Regardless of a brief slippage into zoomorphic imagination, Ljuma's take remains anthropocentric, with its emphasis on humanist values such as happiness. Interestingly, the response of critics to *Honeyland*'s aesthetic and narrative articulations of hive-minded care has been filtered through a predominantly humanistic lens too. Humanism in this context refers first and foremost to an anthropocentric mode of ecological engagement. In social ecologist Stephen R. Kellert's taxonomy of nine major perspectives on nature, it corresponds to a particular form of biophilia (experience and appreciation of nature), alongside the "utilitarian, naturalistic, ecologistic-scientific, aesthetic, symbolic [...], moralistic, dominionistic, and negativistic" perspectives.[18] Kellert argues that the humanistic experience of nature involves those feelings of strong affection, care, and nurturance toward individual elements – such as domestic animals and companion species, which have the capacity to offer reciprocity to humans or reach a relational status as family members – rather than the entirety or interdependency of ecosystems. Focusing on Hatidže's care toward family members and companion species, film critic Amy Taubin interprets the hut sequences as invoking a humanism reminiscent of both Kellert's formulation and of classical art. She states that in scenes showing Hatidže caring for her bed-ridden mother and various animals, lit only by an oil lamp,

> the faces of the two women have the chiaroscuro modeling of Italian Renaissance painting. *Honeyland* is a remarkably beautiful movie, all the more so because the images of Hatidže's world testify to the way she regards the creatures in her care—her mother, her dog, her cat, and the bees—with kindness and endless love.[19]

The framing of the mother, the dog, the cat, and the bees as creatures in Hatidže's care here is striking, as it subtly animalizes Nazife while humanizing the animals, by portraying them all as in need of love, nurturing, and compassion.

In contrast to Taubin's humanism, the directors' references to the Nagoya Protocol and the UN discourse on biodiversity bring the film closest to Kellert's "ecologistic-scientific" category of approach to human-nature relationships. Such ecologism offers an epistemological lens that places "an emphasis on interconnection and interdependence in nature" as well as on its self-organization and complexity, as established by environmental science. Ecologism does not necessarily favor companion species like humanism does, but it can be similarly anthropocentric insofar as it deems self-organization and biodiversity important, primarily for its role in sustaining human life. Therefore, while the ecologistic perspective has affinities with onto-ethological thinking, it does not have the latter perspective's radical approach toward animality.

Lastly, the idea of a likeness between humans and honey bees in the film suggests that the documentary's approach to human-nature relationships is filtered through yet a third lens: the "symbolic." The symbolic highlights an attitude toward nature that locates in human-animal analogies a rich resource for metaphor, in the manner of a cultural poetics. It is possible to interpret *Honeyland*'s human-bee analogy as adhering to common or recognizable patterns of animal, insect, and hive symbolism, instead of an understanding of the two species as channeling human values in a humanistic fashion or an exploration of their eco-systemic roles from an ecologistic perspective.

During a conversation with BBC News' Tamara Kovacevic, co-director Kotevska alludes extensively to the symbolic use of a human-honey bee analogy as plot device in the film. In a section titled "Bees and humans are similar," Kovacevic's article breaks down the three ways, in which the analogy is highlighted in the film. Accordingly, "Hatidže says her bees are uniquely resilient and can survive very high and very low temperature, unlike many other species. The film shows that to be true of the people living in the region too."[20] This is one way that humans and bees are portrayed as similar. Additionally, "In the film, Nazife never leaves the house, but her wisdom guides her daughter at the time of crisis," resembling the relationship between queen bee and worker bees. And lastly, a conflict played out between human neighbors is mirrored by the interactions between competing bee colonies. As Kovacevic notes, "The Sam family that comes later are the other group of bees who are attacking the previous group of bees, which is Hatidže and her family. We really enjoyed making this comparison during the shooting."

This detailed breakdown (imagining the human subjects of the film to be as adaptable, communal, and territorial/colonizing as bees respectively) reflects the film's symbolic framework, recognizing in bees various qualities that help us contemplate human behavior. Whereas the humanistic perspective humanizes bees, the symbolic imagines the humans as bee-like; there is a reversal of the analogy at play, which nonetheless serves an anthropocentric epistemology and sustains human "ipseity" or self-reflection.[21] Although insects have historically been marked as humanity's other, "their swarming, hiving, and colonizing behaviours reproduce various self-identifications in human beings."[22] The three qualities that the documentary distills from bees form the basis for such self-identifications. The bees' adaptability or resilience finds a correlate in humans' ability to live in harsh conditions, exemplified by Hatidže's adaptation to the barren and infrastructure-less landscape of Bekirlijia village as well as the nomadic lifestyle of the Sam family. Bees' swarming behavior is directly linked to the humans' need to change location to adapt to seasonal changes, too. Indeed, the film pays special attention to filming the swarm process, which the crew initially missed and had to wait an entire year to capture. Similarly, the divide between the good hive and the bad hive, represented by the antagonism between Hatidže and the Sams, can be interpreted as demonstrating colonizing behavior in both species.

Notably, *Honeyland* extends the bipolar way that the hive has been historically conceptualized and visualized by the popular imagination. In *Poetics of the Hive*, Christopher Hollingworth takes up representations of the hive figure in literature and states that within its literary evolution, "the Hive usually appears as a special instance of metaphor: a poetic picture" that perceives insect collectivities to be "living emblems of the human social order and our place within it."[23] What the poetics of the hive has produced, over time, is a composite image of social order that carries aspects of both Dante's heavenly bees and Milton's infernal swarm: the angelic and the demonic aspects of it are parts of the same mental structure. If the representation of Hatidže and her hives draws upon the angelic, that of Hussein Sam's "other group of bees who are attacking the previous group of bees, which is Hatidže and her family" (as mentioned in Kotevska's description) draws upon the demonic aspects of the composite image in this regard.

Regardless of the differences between these three compelling perspectives (humanistic, ecologistic, and symbolic), what is noteworthy in their compossibility in *Honeyland* is the exemplification of how "the value of nature documentaries can be approached from diverse epistemological positions."[24] To the above epistemologies, I add a fourth – ethological – to explore where the film's environmental wager ultimately lies.

Toward an ethology: arthropods in modernity

In all of the three competing interpretations mentioned so far, there remains a slippage, a reversibility of the film's human-centered framing. In imagining bees as human-like and humans as bee-like, *Honeyland* occasionally objectifies or arthro-pomorphizes its human subjects, refracting them through the lens of animality. John Mullarkey argues that while our approaches to animals are often "philosomorphic" (reflecting a desire to refract the animal through the image of one's ideas), the Real of animals always mutates or morphs "what counts as philosophy – thought, reason, logic. And cinema – which is a kind of animal thought in all of us – provides just such a mutation."[25] In other words, both animals and cinema present a zoomorphic potential to make us think and see the world differently, not necessarily through representations but through their irreducible materiality. The opening sequence, mentioned above, taps into this zoomorphic potential in that it lets a subject and object-molding encounter unfold between Hatidže and the cliffside colony of bees. In that encounter, Hatidže's biomimicry appears as both a technique inherited from generations of apiarists in her family (Samir Ljuma states that Hatidže told them not to wear dark colors in order to avoid getting stung, which suggests that her outfit is a product of careful thought) and as an openness to letting her singularity be impacted by the subject-shaping powers of the colony and the multitude.[26]

The aforementioned "half for me, half for you" line can be interpreted as pointing to unassimilable animal worlds in a similar way. In a scene eight minutes into the film, Hatidže shares advice carried from her ancestors, as she feeds worker bees some honey: "Hem size hem bize, yari sana yari bana." This line is translated by the subtitles as "one half for you, one half for me," skipping the first part of the sentence, likely because it seems reiterative to a certain extent. Yet while the second part resonates with calls for sustainability, the first part of the sentence is poignant too, indicative of a zoomorphic formulation that imagines singularity and multitude together. The full translation would suggest: "To both you (plural) and us; one half for you (singular), one half for me."

Returning to the opening sequence, part of what blurs the boundaries between the human and arthropod life-worlds in an un-anthropocentric way, is the use of drone imagery (absent from the rest of the film) and the creative soundtrack (also subsumed by the filmmakers' emphasis on visual storytelling later). Laid under the opening sequence is a musical track, composed of non-lexical vocalizations in harmony; the track, titled "Hatidže's Secret," fades with vocals intoning a protracted note, the pulsing reverb of which prepares the viewer for the droning sounds of the colony, about to become audible at the end of the cliffside path. The sound transition supports the idea of an interaction between distinct umwelts à la Uexküll, with human voices mimicking the sounds of the colony and reinforcing Hatidže's own visual biomimicry. While the scene becomes the entry point to the documentary's argument about sustainability as resting upon a few simple protocols (or "secrets"), it also points to what Eric Brown calls, following Kellert, "the radical autonomy" of insects: "they exist beyond our capacity for language, specifically unnameable by anyone but themselves."[27] In the nonverbal vocalizations of the soundtrack, the viewer is presented with the potential of cinema to visualize and sound the unique life-worlds of non-human species, in ways that appear linked to but also irreducible to ours.

While the rest of the film adopts a less experimental aesthetic, it poignantly switches into an observational style that foregrounds visual communication, developed as a response to the language barrier the filmmakers faced while filming. The documentary subjects, the Muratova and Sam families, are all ethnic Turks, speaking in an Ottoman-Turkish dialect that the film crew could not understand; therefore, the documentary was filmed with attention to the relationships between the two families based on the subjects' bodily interactions and visible emotions, rather than dialogue. Kotevska explains:

> When we came to editing, we spent a week or more just being completely stuck. We spent hours just thinking what to do with this material. In the end we came up with the best possible solution and that's to *edit on mute* [emphasis mine]. This gave us the power of the visual story that we made. This was the most valuable lesson for us as the authors of the future films. We are now very trained to see the stories that are created visually.[28]

There is a more overt ethological realism in this aesthetic, in that the interaction between the human and honey bee worlds – the Sams' instrumental approach to animals, Hatidže's appreciation of them as they are, and the indifference of honey bees to human conflicts – is conveyed through a predominantly visual cataloging of bodies, affects, and behavioral patterns. This gives the film the appearance of a study on human-animal-habitat relations or the milieus of behavior, in which different species encounter one another.

The film's audiovisual strategies allow openings into a different understanding of humans and bees then, as occupying distinct life-worlds, subjectivities, and physiologies that refract both species through animality (which is to say, a non-human-centered lens). This grants the film a sense of wonder toward the natural world that is evocative of the spirit of early science films, such as J.C. Bee-Mason's *The Life of the Honey Bee* (1911). Revisiting the tradition, Oliver Gaycken calls popular science films "devices of curiosity," due to the fact that they put aside objectivity's restraint and valorized "engagement with the senses and the affective states of curiosity and wonder" instead.[29] While the aesthetic of *Honeyland* does not draw from "cabinet of curiosity" style display traditions like early science films (Markus Imhoof's 2012 documentary *More than Honey* steers closer to that aesthetic, with forays into macro-photography), it still presents an affective engagement with life-worlds that evokes curiosity.

Thinking about *Honeyland* through an ethological lens also allows a different interpretation of the film's narrative, which can be taken as an indictment of modernity on the surface, with its emphasis on the degradation in biodiversity, habitats, and sustainable cultural traditions. In *Electric Animal*, Akira Lippit states: "No longer a sign of nature's abundance, animals now inspire a sense of panic for the earth's dwindling resources."[30] That has pushed animals to a state of perpetual vanishing in the popular imagination, as the go-to trope in narratives of ecological decline under the impact of modern life. One of the names given to this trope is "animal modernity" or "animals in modernity" vis-à-vis its re-invigoration in eco-cinema.[31] *Honeyland*'s mapping of the anxieties surrounding biodiversity loss onto the images of threatened European honey bees and Balkan beekeeping traditions can be viewed as another eco-cinematic iteration of the trope (an invocation of "arthropods in modernity," perhaps), yet the film's audiovisual strategies frame the human subjects as animals in modernity themselves, once again pointing to a boundary-collapsing parity. On the one hand, this can be taken as reducing both bees and humans like Hatidže to things to be preserved. On the other hand (and I think more importantly), it suggests that both bees and (all) humans deserve treatment as full subjects, with respect toward their life-worlds.

Ultimately, *Honeyland*'s ethological realism should make us question what the ecological wager of the human-bee parity in the film is and what ethics of nature it promotes. This chapter has argued that in combating biodiversity loss, the film urges for a perspectival shift: from human-centered approaches toward the natural world to a framework that highlights the relations and interactions between species, including but not limited to the human. Ethological realism, when understood as reflecting that shift, has the potential to offer unique narrative, aesthetic, and ethical responses to contemporary ecological problems. It also presents a new venue for documentary's crossings with science, which have historically played an important and equally controversial role in its knowledge production (due to pre-climate change era science's claims to transparency, objectivity, and certainty clashing with documentary's desire to critique such claims). Today's environmental science acknowledges that we have to rethink realism, question our certainty about the world, and embrace the fact that the realities that we are going to encounter in the near future will be nothing like what has come before.[32] *Honeyland*'s realism is compelling in this regard too, since its deep respect for environmental science manifests itself not in the pursuit of certainty and objectivity, but in pursuit of an intimate window (through a buzzing crevice, home to a cliffside beehive perhaps) to unstable ecological realities and uncertain futures.

Notes

1 Lynn Badia, Marija Cetinić, and Jeff Diamanti, eds., *Climate Realism: The Aesthetics of Weather and Atmosphere in the Anthropocene* (New York: Routledge, 2021), 4–6.

2 So far, I have only encountered the use of "ethological realism," in a context that might be relevant to this chapter in S. Pearl Brilmyer's *The Science of Character: Human Objecthood and the Ends of Victorian Realism* (Chicago: University of Chicago Press, 2022), which has a chapter on 19th-century novelist Olive Schreiner's ethics of nature. At the time that I wrote this chapter, however, the book was not published yet and only the table of contents was made available on the publisher's site. Therefore, I did not have the opportunity to compare Brilmyer's literary approach to the concept with my eco-cinematic formulation.

3 Thomas D. Seeley, *The Lives of Bees: The Untold Story of the Honey Bee in the Wild* (Princeton, NJ: Princeton University Press, 2019), 291.

4 Charlotte Sleigh, "Inside Out: The Unsettling Nature of Insects," in *Insect Poetics*, ed. Eric C. Brown (Minneapolis: University of Minnesota Press, 2006), 281.

5 Gene E. Robinson, "Darwinian Bee-Keeping: Lessons from the Wild," *Nature* 571 (2019): 34–35.
6 Stephen R. Kellert, "Values and Perceptions of Invertebrates," *Conservation Biology* 7, no. 4 (1993): 845–55. Notably, bee die-offs were also central to the environmental debates in the 20th century, as they led to the birth of the American environmental movement through Rachel Carson's influential book *Silent Spring* (1962), which opens with a discussion of the devastating effects of insecticides.
7 Chris O'Falt, "Light, Cinema, Conflict: The Deceptively Simple Brilliance of *Honeyland*," *IndieWire*, January 6, 2020. https://www.indiewire.com/2020/01/honeyland-tamara-kotevska-ljubo-stefanov-cinematography-editing-1202198905/.
8 Swiss Agency for Development and Cooperation, "Press release: Honeyland Oscar nominee," January 14, 2020.
9 See the United Nations' "Convention of Biological Diversity." Signed at the 1992 Rio Earth Summit, the CBD constitutes the urtext upon which much of the contemporary activities of conservation-focused agencies build. https://www.un.org/en/observances/biological-diversity-day/convention.
10 Darianna Cardilli, "'Honeyland': A Mesmerizing Film on Humans, Nature and Bees," *International Documentary Association*, January 16, 2020.
11 Jakob von Uexküll, *A Foray into the Worlds of Animals and Humans: With a Theory of Meaning* (Minneapolis: University of Minnesota Press, 2010).
12 Brett Buchanan, *Onto-Ethologies: The Animal Environments of Uexkull, Heidegger, Merleau-Ponty, and Deleuze* (Albany: SUNY Press, 2008).
13 Felice Cimatti, "From Ontology to Ethology: Uexküll and Deleuze & Guattari," in *Jakob von Uexküll and Philosophy: Life, Environments, Anthropology*, eds. Kristian Köchy and Francesca Michelini (London: Routledge, 2019), 175.
14 Elena del Rio, *The Grace of Destruction: A Vital Ethology of Extreme Cinemas* (New York: Bloomsbury, 2016), 21.
15 Joshua Malitsky provides a comprehensive summary of the evolution of such investments in his article "Science and Documentary: Unity, Indexicality, Reality," *Journal of Visual Culture* 11, no. 3 (2012): 237–57.
16 "Honeyland: Press Notes," *Honeyland* website, https://honeyland.earth/press/.
17 Documentary Weekly, "Interview with Honeyland Cinematographer Samir Ljuma," *YouTube*, November 15, 2019. https://www.youtube.com/watch?v=hn2VunepVtc.
18 Stephen R. Kellert, "The Biological Basis for Human Values of Nature," in *The Biophilia Hypothesis*, eds. Stephen R. Kellert and O. Edward Wilson (Washington, DC: Island Press, 1993), 44.
19 Amy Taubin, "Swarm and Tender," *Artforum*, July 26 2019. https://www.artforum.com/film/amy-taubin-on-honeyland-2019–80406. Daniel Petrick of Boston Review similarly likens the interior scenes to Caravaggio paintings. See Daniel Petrick, "Untangling Fiction and Reality in the Balkans," *Boston Review*, December 17, 2020.

20 Tamara Kovacevic, "Honeyland: Life Lessons from Europe's Last Wild Beekeeper," *BBC News*, February 9, 2019.
21 Nicole Anderson, "Pre- and Posthuman Animals: The Limits and Possibilities of Animal-Human Relations," in *Posthumous Life: Theorizing Beyond the Post-Human*, eds. Jami Weinstein and Claire Colebrook (New York: Columbia University Press, 2017).
22 Eric C. Brown, "Introduction," in *Insect Poetics*, ed. Eric C. Brown (Minneapolis: University of Minnesota Press, 2006), xi.
23 Cristopher Hollingsworth, *Poetics of the Hive: Insect Metaphor in Literature* (Iowa City: University of Iowa Press, 2001), xi–xii.
24 Adrian Ivakhiv, "Green Film Criticism and Its Futures," *ISLE: Interdisciplinary Studies in Literature and Environment* 15, no. 2 (Summer 200): 4.
25 John Mullarkey, "Animal spirits: Philosomorphism and the Background Revolts of Cinema," *Angelaki: Journal of the Theoretical Humanities* 18, no. 1 (2013): 11–29.
26 What appears genuine in that scene immediately gets co-opted into the film's more restrictive narrative later, of course, as we see Hatidže wearing the same clothes for almost the entire duration of the film (it is plausible that the filmmakers requested it for facilitating continuity).
27 Brown, xii.
28 Kovacevic, "Honeyland: Life Lessons from Europe's Last Wild Beekeeper."
29 Oliver Gaycken, *Devices of Curiosity: Early Cinema and Popular Science* (Oxford: Oxford University Press, 2015), 4.
30 Akira Mizuta Lippit, *Electric Animal: Toward a Rhetoric of Wildlife* (Minneapolis: The University of Minnesota Press, 2000), 1.
31 Belinda Smaill, "Documentary Film and Animal Modernity in *Raw Herring* and *Sweetgrass*," *Australian Humanities Review* 57 (2014): 61–80.
32 This is one of the main arguments of Amitav Ghosh's influential book *The Great Derangement: Climate Change and the Unthinkable* (Chicago, IL: The University of Chicago Press, 2016).

3 "In Europe, no one was paying attention"

Honeyland on the festival circuit

Ilona Hongisto

On January 13, 2020, the North Macedonian film *Honeyland* was nominated for an Academy Award for Best International Feature Film and Best Documentary Feature. This historic accomplishment came a year after the film's successful world premiere at the Sundance Film Festival on January 28, 2019.[1] At Sundance, Tamara Kotekovska and Ljubomir Stefanov's documentary won an impressive three awards: the prestigious World Cinema Grand Jury Prize for Documentary, the World Cinema Documentary Special Jury Award for Impact for Change, and the World Cinema Documentary Special Jury Award for Cinematography. Although the Academy Awards ended up with Bong Joon-ho for *Parasite* and Steven Bognar and Julia Reichert for *American Factory*, respectively, *Honeyland* received numerous other accolades in 2019, such as the Best International Film Award at Docaviv in Israel and the Special Jury Award at CinéDoc Tbilisi in Georgia, to name just a few.

Despite these accomplishments, *Honeyland*'s performance on the festival circuit during the award season was also marked by exclusion from key film festivals and major competitions. Atanas Georgiev, the film's producer and editor, summarizes his sentiment concerning these omissions saying, "In Europe, no one was paying attention."[2] Indeed, whereas the film's opening in North America was extraordinary, its subsequent festival presence in Europe was less pronounced in comparison. In 2019, the film was absent from DocPoint Helsinki, a festival that coincides with Sundance, and the March edition of Denmark's CPH:DOX. It was not included in competition at major European documentary film festivals later in the festival calendar, namely Sheffield Doc/Fest, DOK Leipzig, and IDFA in Amsterdam, and it was not picked up for the A-list festivals in Berlin, Cannes, or Venice.[3] Coming off the success at Sundance, this might seem perplexing.

DOI: 10.4324/9781003124573-4

In what follows, I use Georgiev's off-the-cuff remark to elaborate on the film's seemingly disparate receptions and acknowledgments on the international film festival circuit. His comment, which is easily countered by looking at the number of screenings, audience reception, and sales figures, provides an entry point to a debate on the mechanisms of the festival circuit and the accumulation of attention. I argue that the remark ultimately says more about the circuit itself than about the performance of the film.

The "festival circuit" is an amorphous term that refers to a system of multiple interconnected events. Skadi Loist argues that the composition of the film festival circuit is highly relational, depending on the trajectory of a given film.[4] The festival circuit takes form as films pass through events dispersed on the yearly festival calendar, each film creating its own path and consequently iterating a singular version of the circuit. These iterations are conditioned by the diverse aims, regulations, and hierarchies between different festivals, which complicates the impression of the circuit as a smooth set of connections. Thomas Elsaesser and Marijke De Valck use the term "network" in their respective accounts of the festival system to highlight this complexity of functions, layers, and actors.[5]

The festival circuit is typically distinguished from the commercial circuits of distribution and exhibition and understood to operate, rather, according to the logic of cultural capital. Within the circuit, individual film festivals function as nodes that attribute value to the films they present. De Valck speaks of film festivals as "gateways to cultural legitimization" highlighting the function of festivals in distributing cultural capital.[6] The value attributed at different festivals is, however, not equal in measure as some festivals have more cultural capital than others.[7] The attribution of value happens most explicitly through nominations and awards, but cultural legitimization also happens in the form of "festival buzz."[8] Explicit honors are often preceded or accompanied by rumors, talk, and excitement that form the other facet of attention a film can accrue on the festival circuit. Awards and buzz, then, offer two distinct approaches to contextualizing and evaluating the attention *Honeyland* accumulated on the festival circuit in 2019.

In this essay, my aim is to discuss *Honeyland*'s travels on the festival circuit and to reflect on this journey in relation to cultural legitimization. My treatment contributes to debates on the relational and networked character of the festival circuit by highlighting the gaps in the film's accumulation of attention. Indeed, I argue that the system of interconnected festivals comes with stratified disconnections that play a part in cultural legitimization. In the case of *Honeyland*, this is most explicit in the divide between the North American and European festival circuits that Georgiev refers to in his remark.

The structure of this essay accounts for the understanding that film festivals are much more than mere sites of exhibition. They play a part in production processes and distribution deals and are thus better defined, according to Dina Iordanova, as "industry nodes."[9] I have layered the three intersecting modes of production, exhibition, and distribution into my discussion of *Honeyland* in order to elucidate this complexity of the festival circuit. I begin by mapping the film's production history to analyze its later presence on the festival circuit, then elaborate on festival hierarchies as mechanisms of allocating attention, and finally conclude with a brief discussion of the sales and distribution of the film as they relate to the circuit.

A single-origin documentary in an era of co-productions

Honeyland has its roots in a short documentary film about Lake Prespa in North Macedonia. The same team – Ljubomir Stefanov and Tamara Kotevska as directors, Fejmi Daut and Samir Ljuma as cinematographers, Atanas Georgiev as editor – were commissioned to do a film on the restoration of the lake for the United Nations Development Program (UNDP). This became the short film *Lake of Apples* (2017), which set the tone for what was to become *Honeyland*.

While engaged in the UNDP commissioned project, the team started on a second commission, which Georgiev describes as a "corporate video."[10] This project was to become a short environmental documentary for the Nature Conservation Programme (NCP) in North Macedonia, launched by the Swiss Agency for Development and Cooperation (SDC), focusing specifically on the preservation of the river basin in the Bregalnica region.[11] While researching for the project, the team discovered rocky beehives in the nearby mountains, which led them to Hatidže Muratova, the beekeeper who would later become the protagonist in *Honeyland*. The filmmakers note that meeting Hatidže was a turning point that steered them away from the commissioned short documentary toward a larger film project of which the contours were yet to be defined.[12]

Shooting began in December 2015 with resources from the NCP commission. The initial budget of the corporate video was about €50,000 – funded by SDC and the environmental consultancy company Pharmachem – and it enabled the onset of the larger film project featuring Hatidže. Indeed, the feature documentary was shot with money initially allocated for the short film with the same team and equipment already in place. Both SDC and Pharmachem are acknowledged in the production credits of *Honeyland*, indicating the backstory of the film in international nature conservation programmes.[13]

The cinematographers and the directors spent 100 shooting days over three years producing 400 hours of material in the remote location on the North Macedonian mountains. Hatidže's home was not far from the closest village in kilometers, but the last 20 kilometers was such rugged terrain that most cars would not make it. Due to the challenges in access, the shoots took up to four days at a time, after which batteries needed to be recharged and supplies replenished.[14]

In 2017, after more than a year of shooting, the team was awarded 3,000,000 MKD (roughly €50,000) from the North Macedonia Film Agency for production development.[15] In August 2017, *Honeyland* was featured at the *CineLink Work in Progress* forum at Sarajevo Film Festival, where the filmmakers won the Turkish National Radio and Television Award of €30,000.[16] These funds became available only a year after they had been awarded, which meant that besides the initial support from Pharmachem and SDC, the shooting stage was mainly funded by "enthusiasm and personal credit cards,"[17] which is not uncommon in the documentary world.

One thing that stands out in the film's production history is that Sarajevo's *CineLink Work in Progress* is the only pitching and development forum the project took part in. In addition, the film is a "single-origin" small nation production with no European or international co-production partners. This is striking in a climate where European documentary films on the festival circuit tend to be transnational co-productions almost by default, and the strategy for attracting funds and partners is to attend industry forums organized in conjunction with film festivals.[18] Viktor Kossakovsky's *Gunda* (2020), for instance, was produced by the Norwegian company Sant & Usant Production (Anita Rehoff Larsen) in collaboration with Louverture Films (Joslyn Barnes, US) and Hailstone Films (Charlotte Hailstone, Scotland). It was pitched at IDFA Forum in 2017 and toured the festival circuit to great acclaim in 2020.

When discussing a film's performance on the festival circuit, it is crucial to look at festival presence also in the production stage. Sarajevo's *CineLink* is one in an array of forums where films are developed from ideas to final products through panels, roundtables, and one-on-one meetings with industry experts and stakeholders.[19] This is significant in the context of the present discussion because forum participation may facilitate attention at a later stage in a film's life. The buzz around a film may begin with a successful pitch, and festivals are known to program and foreground films they have supported in the production stage.[20] *Gunda* is a fitting example here, too, as buzz around the film started building after Kossakovsky's artistic performance on the pitching stage at IDFA Forum. The unorthodox pitch by a respected *auteur* left a mark on those present, including myself, generating buzz around the project. Put differently, festivals as industry nodes contribute to the accumulation of attention starting from the production stage. Not attending industry forums takes away this stage in the accumulation process.

In his retrospective account, Georgiev explains that they did not take the project to other pitching forums mainly because of frustrating previous experiences. He says that coming from Eastern Europe, especially Macedonia, soliciting funds or co-production partners at pitching forums is disagreeable because of a general atmosphere in which, he says, "they just don't trust East Europeans."[21] Rada Šešić – founder of Sarajevo's *Docu Rough Cut Boutique* – clarifies this position by pointing out that Balkan directors pitching local projects at international forums have been told more than once by funders that a similar story by a French/German/English filmmaker has already been commissioned. She notes that the industry platform at Sarajevo was designed to respond to this issue and to enable Balkan filmmakers to tell Balkan stories.[22]

In addition to the implication that confidence still lacks between East and West Europeans, Georgiev's account is interesting because it locates the film outside the more common narrative of forum presence and successful production partnerships. Heino Deckert, the agent responsible for *Honeyland*'s world sales, enhances the alternative production narrative by noting that keeping decisions within the small team was fundamental for letting the film brew. As the shoots were spread out over several years, the argument and the perspective of the film took time to crystallize. A production strategy dependent on pitching forums and co-production partners would have required a sharp tagline to begin with, and this could have impeded the tentative, exploratory formation of the film.[23] In Deckert's reflection, not taking part in pitching forums and not soliciting production partners corresponded with the specific cinematic needs of the project.

The local production strategy was complemented by a strategy of applying for film funds to cover specific costs in the budget. Although film funds, too, are typically associated with film festivals, applying for funds to cover specific costs is less performative as applications do not involve pitching the project in front of a live audience. Here, *Honeyland*'s strategy was not selective, and applications were sent to open calls around the world. The one organization that got back to them with positive news was San Francisco Film Fund. In August 2018, the *Honeyland* team was awarded $25,000 USD from the SFFilm Documentary Film Fund, which enabled them to finish the sound mixing of the film. Rana Eid of DB Studios in Beirut was engaged to create the sound design to a finished edit of the visuals. This was a particularly crucial stage in the making of the film, as the lack of a dedicated sound recordist on location had led to recordings of inferior quality. Eid created the sounds of wind blowing and bees buzzing at her studio in Lebanon, finishing the sound design in December 2018.

Georgiev mentions two Directors of Artistic Development at SFFilm Fund by name – Caroline Von Kuhn and Lauren Kushner – and notes that they were the ones who opened the doors to the film in North America.[24] Just before the Sundance nominations were announced, SFFilm Fund advised Georgiev to engage Submarine as the film's North American sales agent. Ben Schwartz at Submarine confirms this and states that the connection was formed in early November 2018, when he was in San Francisco scouting for films. The SFFilm Fund team presented him with a rough-cut of *Honeyland* that was, at that point, "completely under the radar."[25]

Submarine took the film on and started building on the premiere at Sundance. A PR team headed by Ryan Werner of Cinetic Media was established to work the press and to attract distributors to the film at Sundance. The strategy was to hold off from showing the film to anyone before the festival and, rather, to insist on everyone seeing it there. Schwartz describes this as a "you have to see it to understand it" strategy that relies on the audiovisual impact of the film on the audience, leading to word-of-mouth and positive press.[26] After the successful premiere, SFFilm Fund gave the *Honeyland* team a further $50,000 USD for distribution costs, which in the words of Georgiev "saved them" so they could respond to the eruption of interest in the film. This involved establishing a company in the United States to insure the film for North American distribution.

The film's production history contains features that put Georgiev's sentiment of "nobody paying attention" into perspective. Although the *CineLink* forum at Sarajevo was crucial in the development of the film, *Honeyland* evolved outside the major industry forums and funding systems available for European documentary productions.[27] Not being a co-production and not being nurtured by any of the major development forums, the film had no ties that would guarantee its attention on the yearly festival map. In Europe, its original footing was regional, a feature acknowledged with a screening in competition at Sarajevo Film Festival on August 17, 2019.[28]

Interestingly, the disconnect with European co-production procedures and funding tools led to a connection with SFFilm Fund and an alternative route for the film in North America. This is crucial from the point of view of cultural legitimization as premiering a European documentary film in North America may ultimately be more productive. Heino Deckert notes that buzz generated in North America tends to travel back to Europe, whereas buzz generated in Europe rarely finds its way across the Atlantic.[29] Following this line of thought, it is more appealing for a European documentary to open in North America, as success there will lead to attention being paid also in Europe. This pits North America and Europe against each other in a gesture that echoes the history of film festivals: the first festivals were launched in Europe in response to Hollywood's dominant position in distributing value and attention.[30] Contemporary European documentary films enact a reversal of this position in seeking cultural legitimization from North American events. The appeal of this route is demonstrated by the fact that the three European-led productions that generated the most attention in 2019 had their world premieres in North America: *Honeyland* opened at Sundance, *For Sama* (Waad Al-Kateab & Edward Watts) at SXSW, and *The Cave* (Feras Fayyad) in Toronto.[31] All three were also nominated for Best Documentary at the 2020 Oscars.

Festival selection and the premiere paradox

Georgiev explains how he, assured by being nominated for Sundance in November 2018, contacted Berlinale programmers to assert interest in being part of the February 2019 festival. In response, he got an invitation to present the film in the *Culinary Cinema* section of the festival. *Honeyland* ended up not screening at Berlinale and, for Georgiev, the proposed relegation of the film to the *Culinary* section of the program was an insult incited by screening the film at Sundance first.[32]

This concerns the soft politics of film festival programming. Thomas Elsaesser describes film festivals as "gentle gatekeepers" in the business of "supporting, selecting, celebrating and rewarding" the yearly output of films.[33] Given the number of films submitted to festivals each year, it is already an achievement to be included in a festival program.[34] However, the gatekeeping function comes with hierarchies and mechanisms that complexify the celebratory outlook of film festivals. The accumulation of attention is dependent on where a film opens, which festivals it subsequently screens at, and where it is placed in the program.[35] The Berlinale incident puts this aspect of the circuit into relief.

What is crucial here is that leading film festivals assert their power by prioritizing world premieres in their competition sections. Berlinale prioritizes world premieres although international and European premieres can be accepted in some of the less prestigious competition sections. The Official Selection at Cannes rules out prior presence at any other motion picture event. In 2019, Sheffield Doc/Fest listed a UK premiere as the minimum requirement, although priority was given to world premieres. Eligibility rules at both DOK Leipzig and IDFA prioritize world premieres in main competitions.[36] This puts festivals in competition with one another over world premieres. Denominators such as "international premiere" or "European premiere" are used to indicate the less prestigious 'after-the-fact' status of a film in a festival program. Paradoxically, having premiered at Sundance, the North Macedonian film was virtually ineligible for major competitions at European film festivals.

After Sundance, it took two months before *Honeyland* premiered in Europe on April 8, 2019 at Visions du Réel in Switzerland, where it screened in the *Latitudes* section of the festival program.[37] The minimum requirement for selection in this non-competitive section is that the film is a Swiss premiere, which diminishes the prestige value of the otherwise visionary program section.[38] At this time, *Honeyland* was both an international and a European premiere, meaning that it would have been eligible for the *International Feature Film Competition* at Visions du Réel, but it was not selected. In June, the film screened out of competition at Sheffield Doc/Fest in the *New/Hits* section for "powerful new masterpieces selected fresh from the best international film festivals."[39] In early November, *Honeyland* screened out of competition in Germany at DOK Leipzig's *Late Harvest* section for festival favorites and later in the month at the *Best of the Fests* section at IDFA in the Netherlands, where it was also tagged as a Dutch premiere.

Honeyland's inclusion in "best of the fests" sections at major European documentary festivals speaks of a shared acknowledgment of the film's status as one of the key films of the year. However, not being screened at the European A-list festivals in Berlin, Cannes, or Venice contributed to the sentiment Georgiev expresses of being deprived of attention. On the documentary circuit, absence from DocPoint and CPH:DOX early in the year, and then being screened out of competition at Visions du Réel, Sheffield Doc/Fest, DOK Leipzig and IDFA adds to the lack of attention Georgiev references. In his rhetoric, this void aligns with the film being a single-origin work from an East European country that nobody can place on the map.[40] However, a more complex explanation can be found in the regulations that determine a film's trajectory after its world premiere.[41]

Waad Al-Kateab and Edward Watts's *For Sama* is useful to compare to *Honeyland*, as the circuits of these two films both overlap and diverge. *For Sama* had its world premiere in Texas at SXSW on March 11, 2019 and it opened in Europe at Cannes. This European premiere was titled a "Special Screening" – the typical out-of-competition slot for timely documentaries – which apparently bypasses the strict world premiere rule of the Official Selection.[42] The film won the Prix L'Œil d'Or for Best Documentary at Cannes and went on to its UK premiere at Sheffield Doc/Fest, where it also competed for the Grand Jury Award. Later in the year, it won the European Film Academy's award for Best European Documentary, an award *Honeyland* was also nominated for.

Comparing these two films displays alternative versions of the festival circuit.[43] Both films premiered in North America, where *Honeyland* garnered robust prestige and loud buzz. *For Sama*'s selection to Cannes and Sheffield presents a more steadfast European tour and thus a different relational iteration of the festival circuit. It also raises questions about the sanctity of premiere regulations as the film's eligibility to the Official Selection is debatable if one reads the regulations to the letter. *For Sama*'s selection to Sheffield was possible in the framework of the 2019 regulations but would no longer be so under the revised criteria.[44] Whatever the reasons behind

For Sama's comparatively stronger European tour, these details speak of the contingent nature of festival selection that expands from official rules to the specific circumstances of a given year; from the tastes of the programmers to the political atmosphere of the moment. In this light, premiere regulations constitute a mutable organism that conditions the travels of films on the festival circuit while being susceptible to change and exceptions.

"Until IDFA": attention converged

Georgiev's remark about nobody paying attention in Europe ends with the temporal qualifier "until IDFA." *Honeyland* had five out-of-competition screenings at the November 2019 documentary film festival, where the buzz that began at Sundance gained in volume. The film came third in the VPRO IDFA Audience Awards, losing by a mere 387 votes to *For Sama*, which took home the prize with 9,694 votes.[45] *The Cave* took second place. All three screened out of competition because of world premiere regulations, the audience votes balancing the exclusion. The five screenings – one of them in Tuschinski 1, the festival's flagship theater – gave *Honeyland* a notable presence in the festival program. Deckert notes that this is a typical way for film festivals to compensate for the strict premiere regulations for films that already have buzz around them.[46]

The buzz around the film was also enhanced by A.O. Scott's list of best films of the year in *The New York Times*. Scott had already reviewed the film favorably in July 2019, but in December 2019, he named *Honeyland* his top entry of the year, describing it as "nothing less than a found epic, a real-life environmental allegory and, not least, a stinging comedy about the age-old problem of inconsiderate neighbors."[47] Scott's list is a gatekeeping operation comparable to festival selection that, according to both Georgiev and Deckert, played an important role in generating attention for the film both in Europe and in North America.

As IDFA is the last major event in the documentary film festival calendar, it is also the time when North American distribution companies are at work on nomination campaigns for the Academy Awards. *Honeyland*'s Oscar campaign was run by its North American distributor Neon and headed by Ryan Werner, the same publicist who was in charge of the film's PR at Sundance.[48] The campaign included special screenings in Amsterdam, Copenhagen, Berlin, and London to attract European members of the Academy.[49]

These events in late 2019 form a point of convergence between the two seemingly distinct markets. This moment is also one of the rare instances where buzz generated in Europe potentially travels back to North America – in the form of Academy votes. Although the two markets operate according to their own schedules and distinct release windows – when *Honeyland* peaked on the European film festival circuit, it was about to be released on North American streaming platforms – the timing and resonances between IDFA, Scott's "best of" list, and the nomination campaign for the Oscars form a node of attention that bridges the gap.

Beyond the festival circuit

One of the anecdotes circulating about *Honeyland* is that it was the cheapest film ever to end up being nominated for an Academy Award.[50] Its total budget was around €250,000 – with support from such divergent sources as the Swiss Nature Conservation Program, a Macedonian environmental consultancy company, the North Macedonian government, Turkish Radio and Television, and San Francisco Documentary Film Fund. Despite the plethora of international resources, the film is a steadfastly local North Macedonian production.

In my discussion of *Honeyland,* the film's production history leads to a distinction between the North American and European documentary markets. The former appears as an alternative and a potentially more potent path of cultural legitimization for European productions. This complexifies the function of film festivals, initially a European phenomenon founded in response to the American film industry, and calls for more research on the power dynamic between the two, particularly in the context of documentary films.

The *Honeyland* team responded to the divide by engaging two sales agents for the film: the film's North American sales was organized by Submarine whereas Berlin-based Deckert Distribution oversaw world sales.[51] Both companies became involved at the rough-cut stage, in late 2018, and they shared the strategy of going for all-rights deals to ensure the film's theatrical release.[52]

Ben Schwartz at Submarine notes that the film's North American distribution deal was sealed with Neon within a week after Sundance. *Honeyland* ended up accumulating $700,000 USD in gross sales, taking the 20th spot on the 2019 list of box office performances for documentaries in North America. For comparison,

The Cave ranks 71st and *For Sama* 74th, with roughly $45,000 USD each in box office gross sales.[53] *Honeyland's* theatrical release window was scheduled for early 2020 in European territories. Deckert had secured all-rights deals with all European territories, but the film's theatrical release was postponed or even canceled in some areas because of COVID-19. Deckert notes that the Netherlands ended up being the most successful European territory because the local distributor played on the buzz generated at IDFA and put the film in theaters right after the festival.[54] This decision channeled the attention accumulated at the festival to theatrical distribution, and it also benefited from the timing given the lockdowns that ensued in March 2020.

As European territories were preparing for the film's theatrical release window in early 2020, *Honeyland* became available for Hulu subscribers in North America. Neon has an exclusive output deal with the streaming platform, and the online availability of the film was further enhanced with transactional VOD streaming on such platforms as Amazon Prime, GooglePlay, and iTunes.[55] In Europe, some territories responded to the impact of the pandemic with extended broadcasting licenses. For instance, in Norway (NRK), Sweden (SVT) and Finland (YLE), scheduled screenings of *Honeyland* on public broadcasting channels were accompanied with extended online availability of 12 months or more.

Finally, even as *Honeyland* began moving beyond the festival circuit, its life continued to be marked by decisions and possibilities relating to film festivals. In this sense, the film's trajectory confirms the definition of film festivals as industry nodes linking production, exhibition, and distribution together. What my analysis of *Honeyland* has shown is that working one's way through the festival circuit – and consequently accumulating attention – is not simply an issue of making the right contacts, being selected to prestigious programs, and winning awards, but also one of opting out of given production models and being excluded from festivals and competition streams. In this way, cultural legitimization at the festival circuit concerns navigating disconnections as much as establishing a relational network of value-adding connections.

Notes

1 *Honeyland* was the first film ever to be nominated for both of these awards. *Collective* (Alexander Nanau, Romania, 2019) repeated the feat in 2021.

2 Interview with Georgiev, 2020.
3 A-list festivals are FIAPF accredited festivals that form the top layer in the international festival hierarchy. They are followed by smaller and specialized film festivals, such as documentary film festivals, that are acknowledged but rank lower in prestige. See: http://www.fiapf.org (accessed April 21, 2021).
4 Skadi Loist, "The Film Festival Circuit: Networks, Hierarchies, and Circulation," in *Film Festivals: History, Theory, Method, Practice*, eds. Marijke De Valck, Brendan Kredell and Skadi Loist (London: Routledge, 2016), 49–50. See also Dina Iordanova, "The Film Festival Circuit," in *Film Festival Yearbook 1: The Festival Circuit*, eds. Dina Iordanova and Ragan Rhyne (St Andrews: St Andrews Film Studies with College Gate Press, 2009), 23–39.
5 See Thomas Elsaesser, "Film Festival Networks: The New Topographies of Cinema in Europe," in *European Cinema: Face to Face with Hollywood* (Amsterdam: Amsterdam University Press, 2005), 82–107 and Marijke de Valck, *Film Festivals: From European Geopolitics to Global Cinephilia* (Amsterdam: Amsterdam University Press, 2007). Both refer to sociological systems and network theory to account for the intricacies of the film festival circuit. Elsaesser reflects on the emergence of the festival network as a response to the studio system in Hollywood. De Valck accounts for the networked structure from the perspectives of European geopolitics and cinephilia.
6 De Valck, *Film Festivals*, 38.
7 Value creation is a key thread in film festival studies, where it has often been addressed through sociologist Pierre Bourdieu's notion of the "cultural field." On the hierarchies of film festivals and value creation, see for example Marijke de Valck, "Fostering Art, Adding Value, Cultivating Taste: Film Festivals as Sites of Cultural Production," in *Film Festivals: History, Theory, Method, Practice*, eds. Marijke De Valck, Brendan Kredell and Skadi Loist (London: Routledge, 2016), 100–16; Marijke de Valck and Mimi Soeteman, "'And the Winner Is …' What Happens Behind the Scenes of Film Festival Competitions," *International Journal of Cultural Studies* 13, no. 3 (2010): 290–307, https://doi.org/10.1177/1367877909359735; Diane Burgess, "Capturing Film Festival Buzz: The Methodological Dilemma of Measuring Symbolic Value," *Necsus – European Journal of Media Studies* 9, no. 2 (2020): 225–47, https://doi.org/10.25969/mediarep/15318; Aida Vallejo, "Rethinking the Canon: The Role of Film Festivals in Shaping Film History," *Studies in European Cinema* 17, no. 2 (2020): 155–69, https://doi.org/10.1080/17411548.2020.1765631; Loist, "The Film Festival Circuit"; de Valck, *Film Festivals*; Elsaesser, "Film Festival Networks."
8 Andrew Sarris first described festival buzz as a contagious 'cinematic virus' in his analysis of Cannes in the late 1970s. Diane Burgess elaborates on the methodological challenges of festival buzz in the film festival value chain, evaluating the ways in which value creation and buzz as attention converge. See Andrew Sarris, "Catch as Catch Cannes: The Moles and the Moths," *The Village Voice*, 12 June 1978, 39–40; Burgess, "Capturing Film Festival Buzz."

9 Dina Iordanova, "The Film Festival as an Industry Node," *Media Industries Journal* 1, no. 3 (2015): 7–11. See also Mark Peranson, "First You Get the Power, Then You Get the Money: Two Models of Film Festivals," *Cineaste* 33, no. 3 (2008): 37–43; Samara Chadwick, "Building Networks: An Interview with Sandra J. Ruch, Director Emeritus of the International Documentary Association," in *Documentary Film Festivals: Changes, Challenges, Professional Perspectives (Vol. 2)*, eds. Aida Vallejo and Ezra Winton (London: Palgrave, 2020), 167–74; Tamara L. Falicov, "The 'Festival Film': Film Festival Funds as Cultural Intermediaries," in *Film Festivals: History, Theory, Method, Practice*, eds. Marijke De Valck, Brendan Kredell and Skadi Loist (London: Routledge, 2016), 209–29; Aida Vallejo, "Defining Documentary in the Festival Circuit: A Conversation with Bill Nichols," in *Documentary Film Festivals: Methods, History, Practice (Vol. 1)*, eds. Aida Vallejo and Ezra Winton (London: Palgrave, 2020), 19–28.

10 Interview with Georgiev, 2020.

11 https://farmahem.com.mk/news/view/84?l=eng (accessed November 19, 2020); https://www.eda.admin.ch/countries/north-macedonia/en/home/international-cooperation/projects.html/content/dezaprojects/SDC/en/2010/7F06872/phase1.html?oldPagePath=/content/countries/north-macedonia/en/home/internationale-zusammenarbeit/projekte.html (accessed April 21, 2021); https://www.eda.admin.ch/deza/en/home/suche/suchresultat.html/content/deza/en/meta/news/2020/1/14/honeyland (accessed November 19, 2020).

12 See for example Daniel Eagan. ""It's a Choice: To Be a Filmmaker or a Human": Ljubomir Stefanov and Tamara Kotevska on Honeyland," *Filmmaker Magazine*, August 22, 2019, https://filmmakermagazine.com/107983-its-a-choice-to-be-a-filmmaker-or-a-human-ljubomir-stefanov-and-tamara-kotevska-on-honeyland/#.YIfCOH0zbvW; Film at Lincoln Center, "Tamara Kotevska and Ljubomir Stefanov on Honeyland, Family & Capturing Macedonia (NDNF19)," YouTube, April 24, 2019, https://youtu.be/_ES4mfYYsII; ICA, "Frames of Representation 2019: Honeyland Intro + Q&A," YouTube, April 27, 2019.

13 The "corporate videos" produced for NCP are available at: https://www.youtube.com/c/NCPBregalnica/videos (accessed April 21, 2021). Authors of the videos are not specified, but some of the videos published around 2014–16 bear stylistic similarities to *Honeyland*. Georgiev confirms that they indeed look like the ones he edited for the commission. Georgiev, personal communication, April 23, 2021.

14 Interview with Georgiev 2020; Will Tizard, "Honeyland DP on Low-Fi Shooting with High-Powered Storytelling," *Variety*, November 25, 2019, https://variety.com/2019/film/festivals/honeyland-camerimage-fejmi-daut-1203401810/; 360 Stepeni, "Finding the Star of Honeyland," YouTube, January 23, 2020, https://youtu.be/dj9mT91fe64.

15 http://filmfund.gov.mk/?page_id=19179&lang=en (accessed November 20, 2020).

16 Georgiev notes that support from Turkish National Radio and Television was not surprising, given that Hatidže is Turkish and speaks a Turkish dialect in the film. The award enabled them to offer support

to their protagonists: they bought Hatidže a house closer to the village and her brother, and the Hussein family got a car. Interview with Georgiev, 2020. There is much more to be said about the ethical choices and decisions that relate to supporting the subjects of one's film financially or materially, but that discussion in beyond the scope of this essay. The directors of the film touch upon some of these ethical dilemmas in Eagan, "It's a Choice."

17 Interview with Georgiev, 2020.

18 Co-productions have become a fixture in the European film field through designated program and funding tools – such as the EU's MEDIA program, Eurimages and Creative Europe – which channel support to transnational collaborations. See Julia Hammett-Jamart, Petar Mitric, and Eva Novrup Redvall eds., *European Film and Television Co-Production: Policy and Practice* (London: Palgrave Macmillan, 2018).

19 On industry forums at documentary film festivals, see Aida Vallejo, "Industry Sections: Documentary Film Festivals between Production and Distribution," *Iluminace* 26, no. 1 (2014): 65–82; Aida Vallejo, "Idfa's Industry Model: Fostering Global Documentary Production and Distribution" in *Documentary Film Festivals: Changes, Challenges, Professional Perspectives (Vol. 2)*, eds. Aida Vallejo and Ezra Winton (London: Palgrave, 2020), 23–54; Ilona Hongisto, Kaisu Hynnä-Granberg, and Annu Suvanto, "The Invention of Northeastern Europe: The Geopolitics of Programming at Documentary Film Festivals," in *Documentary Film Festivals: Changes, Challenges, Professional Perspectives (Vol. 2)*, eds. Aida Vallejo and Ezra Winton (London: Palgrave Macmillan, 2020), 73–91. On the specificities of Sarajevo's industry events, see Jennifer M.J. O'Connell and Annelies van Noortwijk, "Selecting Films for Festivals and Documentary Funds: An Interview with Independent Film Programmer and Advisor Rada Šešić," in *Documentary Film Festivals: Changes, Challenges, Professional Perspectives (Vol. 2)*, eds. Aida Vallejo and Ezra Winton (London: Palgrave Macmillan, 2020), 175–84.

20 See O'Connell and van Noortwijk, "Selecting Films," 181.

21 Interview with Georgiev, 2020.

22 O'Connell and van Noortwijk, "Selecting Films," 181.

23 Interview with Deckert, 2021. This relates to the team being relatively unknown. A filmmaker like Viktor Kossakovsky can accumulate funds, co-producers, and partnerships without having to negotiate their vision, but less established filmmakers seldom have such power.

24 Neither Kushner nor Von Kuhn work at SFFilm Fund anymore; Kushner co-founded Early Bird Films and Von Kuhn is Director of Catalyst at Sundance Film Institute.

25 Interview with Schwartz, 2021.

26 Interview with Schwartz, 2021. Positive reviews came out of the festival, and they were followed by several "making-of" articles that explain how the film was made and how the team met the protagonist.

27 IDFA Forum is the biggest yearly event in the industry calendar with events at Sheffield Doc/Fest, DOK Leipzig, Visions du Réel and CPH:DOX following suit. See Vallejo, "Industry Sections"; Hongisto et al, "Invention."

28 Sarajevo Film Festival is known for its mission to promote regional films. See, for example, festival director Elma Tataragić's introduction (2019, p. 27) in the catalogue of the 2019 festival edition: https://issuu.com/sarajevofilmfestival/docs/25thsff_catalogue_webl (accessed April 26, 2021).

29 Interview with Deckert, 2021. He is also careful to note that success in North America depends on the film. In general terms, story-driven documentaries tend to do well in that market. Ben Schwartz from Submarine confirms this and notes that, whereas European fiction films can be sold with reference to European buzz, the North American documentary market operates independently. Interview with Schwartz, 2021.

30 See, for example, Elsaesser, "Film Festival Networks"; de Valck, "Film Festivals."

31 *For Sama* is produced by Channel 4 News/ITN Productions in the UK for Channel 4 and Frontline PBS in the US. Its "Europeanness" is of course up to debate, but the connection is there. *The Cave* is produced by Danish Documentary with several international co-production partners. Both films are also, of course, Syrian, but they are produced by European companies.

32 Interview with Georgiev, 2020.

33 Elsaesser, "Film Festival Networks," 96.

34 De Valck and Soeteman, "And the Winner is…," 293, 295–97.

35 Deckert points out that opening at an A-list festival such as Sundance, Cannes, or Berlin is better for documentary films from the sales point of view as distributors attend these festivals. They are less likely to attend the prestigious documentary film festivals in Amsterdam, Leipzig, or Nyon. Interview with Deckert, 2021.

36 The specific guidelines for each festival can be accessed through these links (accessed October 5, 2020). They are to the most recent editions of each event, but to my knowledge only Sheffield Doc/Fest has changed their eligibility rules since 2019. The Sheffield regulations listed below are applicable to *Honeyland*. VDR: https://www.visionsdureel.ch/en/festival/sections; Sheffield (2018–2019): https://tinyurl.com/yw6a5rmx; DOK Leipzig: https://www.dok-leipzig.de/en/submission-regulations; IDFA: https://www.idfa.nl/en/info/festival-entry-regulations-2021; Berlinale: https://www.berlinale.de/en/film-entry/guidelines/general-guidelines.html; Cannes: https://www.festival-cannes.com/en/infos-communiques/info/articles/submission-of-films-for-2021.

37 The film's screenings in 2019 have been documented on its official webpage: https://honeyland.earth/screenings/ (accessed December 11, 2020).

38 Among the ten films in the program, nine were Swiss premieres, and *Honeyland* was the only European premiere.

39 Sheffield Doc/Fest Programme Guide (6–11 June 2019), p. 83: https://issuu.com/sheffielddocfest/docs/docfest_programme_guide_2019_issuu (accessed November 12, 2020)

40 Interview with Georgiev, 2020.

41 *Honeyland* was nominated for and won awards at smaller festivals – such as the Polish Docs for Gravity and the Spanish DocsBarcelona – that do not regulate inclusion based on premiere status.

42 I use the term "apparently" as this is a paradox that is not explicated in the eligibility rules for the Official Selection. Special Screenings are part of the Official Selection, and the rules to the Official Selection rule out presence at any other international event (see fn 36). On the presence of documentaries at Cannes, see Eulàlia Iglesias, "Documentaries at the Cannes International Film Festival: Fahrenheit 9/11 and Beyond," in *Documentary Film Festivals: Changes, Challenges, Professional Perspectives (Vol. 2)*, eds. Aida Vallejo and Ezra Winton (London: Palgrave Macmillan, 2020), 113–30.

43 Yet another interesting relational example is Talal Derki's *Of Fathers and Sons* (2017), which was selected for the *IDFA Competition for Feature-Length Documentary* in the November 2017 edition of the festival and two months later won the *World Cinema Grand Jury Prize for Documentary* at Sundance in 2018. Competing at both IDFA and Sundance was technically possible, because IDFA's world premiere requirement was fulfilled and the 2018 edition of Sundance's main documentary competition accepted North American premieres. (Of the 12 films in competition, 10 were world premieres and 2 were North American premieres.)

44 Deputy festival director Melanie Iredale confirms that *For Sama's* inclusion in competition was possible within the 2019 regulations but would no longer be so. Melanie Iredale, Personal communication. April 16, 2021. *Honeyland* premiered in the UK on April 13, 2019, at the Frames of Representation festival at London's ICA. This is just five days after its European premiere in Switzerland.

45 https://www.idfa.nl/en/article/131298/for-sama-wins-the-vpro-idfa-audience-award#:~:text=For%20Sama%20(Waad%20al%2DKateab,winning%20film%20in%20Path%C3%A9%20Tuschinski (Accessed November 18, 2020).

46 Interview with Deckert, 2021.

47 A.O. Scott, "'Honeyland' Review: The Sting and the Sweetness," *The New York Times*, July 25, 2019, https://www.nytimes.com/2019/07/25/movies/honeyland-review.html; A.O. Scott and Manohla Dargis, "Best Films of 2019," *The New York Times*, December 4, 2019, https://www.nytimes.com/2019/12/04/movies/best-films.html.

48 Interview with Schwartz, 2021.

49 Interview with Georgiev, 2020; Interview with Deckert, 2021.

50 Interview with Georgiev, 2020.

51 Submarine has a track record of several award winners from previous years and Deckert Distribution is known for their commitment to Eastern European productions with a strong authorial touch.

52 Interview with Deckert, 2021; Interview with Schwartz, 2021.

53 See https://www.the-numbers.com/market/2019/genre/Documentary (accessed April 3, 2021).

54 Interview with Deckert, 2021. Deckert notes that the film ended up selling to around 60 territories worldwide. He sees this as an optimal and an unusual result for a documentary film.

55 Interview with Schwartz, 2021.

4 Observational time zones

The ethics of *Honeyland*

Linnéa Hussein

Many critics have used the words "beauty" and "beautiful" to describe Ljubomir Stefanov and Tamara Kotevska's 2019 observational documentary *Honeyland.* A.O. Scott, in a review for the *New York Times*, for example, refers to the documentary's opening scenes as "sublime and strange and full of human and natural beauty."[1] Likewise, David Sims from *The Atlantic* comments on the film's "striking natural beauty,"[2] while David Ehrlich from *IndieWire* says the film is a "mesmerically beautiful documentary,"[3] and Michael O'Sullivan, writing for the *Washington Post*, calls it a "surprisingly beautiful little movie."[4] On the surface, the documentary's tale of one of Macedonia's last-surviving rural beekeepers and the ecological and human threats she must face in order to preserve an ancient way of life can indeed present an enticing look. The makers film the beekeeper Hatidže, their main subject, in gorgeous long shots, so that she appears as one with the endlessly magnificent natural landscape. The idyllic representation of the beekeeper, however, contrasts with the representation of her neighbors, the Sams, who are presented as rubes struggling with life under dirty and poor living conditions. This chapter focuses on the less "beautiful" side of *Honeyland*: the representation of the Sams, and, particularly their offspring, who are predominantly shown experiencing chaos, violence, and pain in the film. By studying the portrayal of the Sams in contrast to the "beautiful" depiction of Hatidže, we can see striking differences in how the filmmakers deploy their observational camera depending on *who* they are observing.

Honeyland's filmmakers use the observational camera to create a dichotomy between Hatidže – who appears to care for the land, nature, and her bees – and the Sams – who are shown as lacking care for the land, the animals, and even for their own children. More to the point, different filmmaking choices throughout the documentary make visible how the filmmakers' use – or abuse – of

DOI: 10.4324/9781003124573-5

observational cinema itself manifests as a lack of care, especially for the children's wellbeing. Falling in between the cracks set up by the binary filmmaking protocols that frame Hatidže as "good" and the Sam parents as "bad," the children become part of the mise-en-scène, filmed to lend tone to *Honeyland*'s scenery but not necessarily as three-dimensional subjects worthy of care themselves. In what follows, I examine the treatment of *Honeyland*'s children to interrogate the ethical problems that arise when observational cinema is used as a stylistic choice but is divested of a clear sense of responsibility toward the film subjects. In divorcing the observational camera from its original political ethos, I argue, standards of ethical filmmaking are abandoned, and a problematic instantiation of observational cinema emerges, one that particularly endangers those who cannot fully give their consent to the filmmaking process. Moreover, I suggest that the different ways in which the filmmakers film Hatidže and the Sams have implications for what it means to "observe" in the context of observational cinema.

The mise-en-scène of *Honeyland*'s "time zones"

In works of fiction, the mise-en-scène is carefully constructed to support the characters and narrative. Similarly, in documentaries the social environment of the nonfictional performer may be carefully selected to support his or her signification within the larger documentary context. Often, documentary setting functions to characterize documentary subjects beyond their immediate attributes.[5] In the case of *Honeyland*, however, the human characters are primarily utilized to shape our reading of their environment. As the name of the documentary suggests, the filmmakers' main interest lies with the land; for the most part, its inhabitants are placed within the mise-en-scène as scenery rather than as players. In particular, its smallest human inhabitants – the Sam children – are employed primarily to illuminate Hatidže's and their parents' respective meanings within the documentary's larger narrative context. With no agency of their own, the Sam children are thus used as background to highlight two sets of figures: first, Hatidže as a representative of the land's idyllic timeless past and, second, their parents as markers of the terrifying present that threatens to destroy the land of honey. Paying special attention to *Honeyland*'s elements of film form – its use of narrative design, editing, lighting, and dialogue – uncovers how, despite an observational and ostensibly "neutral" approach, the filmmakers' formal choices present a good versus evil divide that ideologically signifies a value preference for old over new, pre-modern versus modern, and rural versus industrial.

Another way we might think of this is that Hatidže and the Sam parents appear to inhabit different "time zones" within the film.[6] *Honeyland* begins with timeless images of a part of the world seemingly untouched by human intervention. The directors' statement describes the setting as "an unearthly lan [*sic*], unattached to a specific time and geography."[7] Without seeing any indicators of modernity, we follow the beekeeper Hatidže as she climbs the rural mountains to check on her honey, a shot reminiscent of how Brian Winston describes the problematic origins of documentary with Robert Flaherty's "individual heroic 'Inuit,' against the sky in a long shot."[8] Surrounded by nature and dressed in old-fashioned clothes, Hatidže and her mother are subsequently filmed by candlelight as they say "good night" to one another in their modest but cozy house. The first few minutes of the documentary establish Hatidže's life as harsh yet idyllic, as if this woman stands as a rock persevering against modern times. When we follow her to the market in the nearest city, Hatidže's alleged out-of-timeness is further emphasized. Compared to the other shoppers in the market, she seems like a relic from another era; marked by her comparatively old-fashioned clothing, her crooked teeth, and an overall disconnect from modern technological devices such as cellphones, she is temporally set apart from the other people in the city.

This idyllic world-out-of-time is disrupted the minute the Sams and their eight children move in next door. Their arrival signifies the beginning of a different temporality in Hatidže's neighborhood, an intrusion of the more modern world in her agrarian idyll. The Sams are visually and sonically aligned with industrial modernity, and an immediate shift of tone signifying this temporality becomes noticeable in their introductory scene: not only is the sky suddenly grayer, but the soundtrack amplifies the rusty motor of the Sams' vehicle. Their loud truck, tied to an old RV, punctuates Hatidže's romantic landscape with dust and loud noises. We see their arrival through Hatidže's wary eyes, and this sense that the Sams' arrival is unwelcome is further reinforced through lighting. Whereas Hatidže and her mother are filmed in golden rays of sun or candlelight, the Sams are filmed against clouds or gray tones of dust. The golden light and quiet of the sunset are now interrupted by Hussein Sam's hammering of a nail into his car, yet another sign marking the belated arrival of industrialization in Hatidže's tranquil time capsule. Indeed, the scenes featuring the Sams throughout the film are dominated by visual and sonic chaos. Although the film presents itself as observational, with this mode's promise of neutrality, the sound mix and lighting – chosen, albeit natural – assert a clear narrative preference for Hatidže's way of life.

Editing and dialogue further this sense of temporal divide and implicit preference. When the RV comes to a halt, we see Hussein Sam and his sons trying to set up an improvised awning, while at the bottom of the frame a toddler girl almost gets her head stuck in a rusty part of what once constituted a fence. A cut back to a struggling Hussein and his wife suggests they are rubes who cannot find the right tools to remove an old nail from their car. Next, the camera gazes through the open RV door, where free roaming chickens wander about crowded dirty furniture, right next to another little girl. The film then cuts back to Hatidže, calmly telling her mother, "They're Turkish," a statement that lets a viewer unfamiliar with local ethnic animosities ponder what it could mean to be "Turkish" in this context. Earlier in the film, we had witnessed Hatidže lamenting the fact that she and her mother were the only ethnic Turks left in their village. However, the arrival of the Sams does not seem to bring about joy of reconnecting with "her people." The worried look on Hatidže's face seems to imply territorialism more than feelings of anticipated neighborly harmony. The Sams' modernity is framed as overshadowing any potential ethnic alliance.

Key to the characterization of the Sams as antagonistic and contemptible people is the framing of their children. When the RV appears, children of all ages seem to be everywhere. Like the rusty car and the broken nail, their running and screaming suggest an atmosphere of chaos around the Sams. When they sit around Hatidže's backyard to feed a calf, however, they seem to enter her temporality, enjoying the sunny tranquility of her idyll. The children are thus able to travel between their parents' and Hatidže's time zones as their states of chaos or calm are used to characterize the adults' respective positions within the film. Because of the observational style, we do not receive contextualizing information – for the most part, the children go unnamed – nor do we ever witness a conversation between the Sams and their offspring, which might have lent the children a stronger subjectivity of their own. As it is, however, the children function more like props in the mise-en-scène than as actual human beings (Figures 4.1a and 4.1b).

Figure 4.1 The Sam children travel between their parents' and Hatidže's "time zones."

A genuinely ethical documentary filmmaking practice requires that the filmmakers see and experience themselves as co-present, both spatially and temporally, with their subjects. However, by framing Hatidže as timeless and the Sams as dated, the filmmakers remove themselves from either "time zone" and thereby abdicate their responsibility. Indeed, although neglecting to give voice to the Sams' perspectives may do them disservice, the greater ethical transgression of the film is rooted in the filmmakers' refusal to move beyond observation even when a child appears to be in danger. Instead, the representation of violence around and towards children becomes a rhetorical tool that further highlights the dichotomies the filmmakers seek to construct. In addition to the contrast between timelessness and modernity, idyll and dystopia, the filmmakers also emphasize care and its lack as critical value systems that divide the neighbors. Hatidže, as we know from beautiful shots that frame her as one with her environment, cares for her bees, for her mother, and for the land. The Sams, by comparison, do not seem to take good care of anything in their lives; besides being pictured as bad neighbors, bad farmers, and bad beekeepers, they are also presented as bad parents. In an early scene in which Mrs. Sam and the kids are trying to take care of their cattle, we see one of the little boys getting kicked in the chest by a cow. As the scene continues, another son gets kicked by a cow when trying to milk it; the mother yells and runs towards a misbehaving cow as the youngest girl (who is maybe one or two years old) is left sitting on the ground among the free-running, ill-tempered cattle. Here, one can see that *Honeyland*'s children exist within the film to evidence a lack of care; they are there to frame their modern parents as incapable of operating in harmony with their environment. By showing how carelessly the Sams treat their children, animals, and the land, the film seems to suggest that they are not as morally deserving as the timeless, caring Hatidže. In this way, the filmmakers position all of *Honeyland*'s inhabitants as agents of their own prosperity, ignoring the lack of governmental support and the like for the region.[9] The government's neglect is not as easily imaged as that of the Sam parents, and its discussion would complicate the temporal structure established by the film.

Thus, the temporal trichotomy in which each party – Hatidže, the Sams, and the filmmakers – is positioned, as if existing in a different time zone, functions to make it seem as if the filmmakers could not possibly intervene when children get hurt without "time traveling" into the past. Moreover, framing the children as props in the mise-en-scène rather than subjects with narrative agency enables the filmmakers' own lack of care towards these children. The filmmakers' observational "ethics" seem to extend to allowing children to be exposed to danger, which opens up the larger question of their ethical responsibilities as observational filmmakers. Given that the notion of objective observation has been critiqued since the earliest days of observational cinema, what – if anything – is gained by this lack of intervention?

Observational aesthetics, differential ethics

In the late 1950s, technological advances in camera and sound equipment presented new ways for filmmakers to conceptualize documentary films. Coinciding with an awakened interest in humanism following the atrocities of World War II, these new cinematic developments offered filmmakers ways to rethink the ethnographic encounter, moving away from the authoritative expository tradition and towards a desire to let people represent themselves.[10] Early ethnographic filmmakers like Robert Flaherty overtly staged scenes to frame other people according to their imagination.[11] In the 1940s and 1950s, ethnographers like Margaret Mead and Gregory Bateson moved away from staging "the perfect encounter" but nevertheless framed what was filmed with their own voiceovers, often assessing their subjects according to norms of Western civilization.[12] Observational cinema, as the new movement in ethnographic film came to be called, offered filmmakers a way to *observe* without overt intervention through voiceover, intertitles, or other explicit commentaries.[13] This meant filming without speaking to the subjects on camera or otherwise participating in the scene.

Beyond its different aesthetics, early ethnographic observational cinema was also defined by a concern with the ethics of filmmaking. Observation is not synonymous with objectivity, a common misunderstanding. Instead, as Colin Young notes, it "hinged upon the forging of close, personal relationships with subjects."[14] According to practitioners such as David MacDougall and Judith MacDougall, a close relationship between filmmaker and documentary subject needed to be developed before any kind of ethical observation could take place.[15] In ethnographic observational cinema, the duration of events, for example, is an important factor for representing the lived historical world. Instead of editing footage for flow or entertainment, filmmakers like the MacDougalls believed that fidelity could be better achieved by letting the camera witness the amount of real time a conversation or an event would take.[16] Based on mutual trust, observational cinema is intended to operate by establishing certain protocols between maker and subject before filming takes place.[17] One can see a similar approach to ethics in observational cinema's more popular form; direct cinema, as it came to be known in the United States in the 1960s, proliferated through the works of filmmakers like Frederick Wiseman, D.A. Pennebaker, and Robert Drew. While the MacDougalls came from an anthropological tradition that used the camera as a tool to present ethnographic research, the direct cinema filmmakers were more interested in "real people (...) filmed in uncontrolled situations," that is, documenting institutions, events, or performers in the moment rather than extracting ethnographic information about a group of people.[18]

Honeyland does not fit neatly into either of these understandings of observational cinema. Looking at the film, we are seemingly confronted with two different sets of filmmaking protocols, depending on who is in frame. Hatidže, as the film's main subject, is filmed in long shots, oftentimes against a seemingly endless and gorgeous landscape. Her scenes are shot with a calm and steady camera, framing her in picturesque portraits aligned with the ethnographic approach. While there is no overt violation of the norms of the observational mode, the framing, mise-en-scène, and tempo of filming all emphasize and elevate the beautiful aspects of what is there to be observed. The Sams, on the contrary, are filmed in a way that is reminiscent of the direct cinema tradition: hand-held cameras, spontaneous action, and shaky imagery blur the landscape and foreground the human chaos of the moment. The controlled ethnographic composition of Hatidže's beautiful shots give her scenes, including those in which she speaks, a calm sense of authority over the land that contrasts with the seemingly uncontrolled direct cinema approach that is used to characterize – and judge – the Sams as threatening to the natural order. As a consequence of these shifting protocols, we witness two very different types of observational ethics within the same film: one collaborative and the other objectifying.

However, it is the Sams' children who make these different filmmaking protocols most visible. As underaged documentary subjects, they are not in a position to grant informed consent. Their depiction hinges upon the differential relationships established between the filmmakers and the adults, which fluctuate between the collaborative and the objectifying. When the filmmakers do not intervene in those scenes where violence is inflicted upon children, the pivot from the empathetic or romantic look deployed toward Hatidže to an "objective" and objectifying clinical observation is jarring. Given these two starkly different filmmaking protocols, one could say that the children fall victim to the rhetorical dichotomies of the text. They become scenery, necessary to the film's contrasting representation of time, care, and neglect.

Language is one way in which contemporaneity between filmmaker and film subject can be established. Observational cinema as a whole has historically been concerned not only with looking but also with listening. As Grimshaw and Ravitz write, "no longer dominated by the single voice of an anonymous narrator, the observational film brought into focus a range of different voices, ways of speaking, and contexts of speaking."[19] In multiple interviews, Stefanov and Kotevska have admitted to not speaking Turkish, the language of their protagonists. Furthermore, they celebrate this fact to pride themselves in having instead created "a visual narrative."[20] As a result, they privilege the visual over the linguistic in *Honeyland.* Although this may serve the ends of making a "beautiful" film, this elevation of the visual neglects what their subjects might have to say. This decision suggests that the filmmakers' *view* – an outsiders' view – ought to be privileged over the filmed subjects' lived experience. In other words, the desire to tell a purely visual story about Hatidže's life implies an active disinterest in the documentary subjects' own language and rhetoric. Moreover, the fact that the filmmakers did not speak the subject's language makes one wonder how far consent, understanding, or general trust was developed between the filmmakers and subjects.

In addition, the emphasis on the visual rather than linguistic gives the film the feel of a nature documentary. A downside of positioning *Honeyland*'s inhabitants as part of the landscape is that they may be perceived as more animal than human, more metaphoric than material. This zoological and metaphoric treatment is most pronounced in relation to the children. There are few translations of the interactions with or among the children. As a result, even the ones old enough to speak cannot be understood by any viewer who does not speak Turkish. This lack of linguistic agency further reduces them to objects. Abuse becomes naturalized in the name of aesthetics, seemingly justifying the filmmakers' lack of intervention when the children are in danger or hurt. Indeed, the Sams' children receive the same amount of care from the filmmakers as do the cattle or bees. Had the filmmakers understood their speech, their sense of co-presence and responsibility within the scene might have been greater. Without the benefit of shared language, however, the temporal divide between filmmakers and subjects remained in place, allowing for an ethical aporia to emerge.

Observation as cover for neglect

For Jonathan Crary, "observation" as a process goes beyond spectatorship, or onlooking. Depending on context, Crary points out, "to observe" can either mean looking or following a set of rules.[21] Both meanings come into play in the process of making an observational film, in the forms of observing people and observing ethics. Grimshaw and Ravetz effectively apply Crary's definition of observation to observational cinema, explaining it as a "skilled practice that hinges upon a particular kind of active and disciplined engagement with the world."[22] Observation, in other words, is a care*ful* act of engaging with the subjects and thus stands in opposition to a care*less* act, or, neglect, which is usually defined as a failure to care for something. In relation to children, UNICEF includes in its definition of neglect "the failure to properly supervise and protect children from harm as much as is feasible."[23] If the observational camera *fails* to care in the sense of "properly supervise and protect children from harm as much as is feasible," does observation become a form of – or excuse for – neglect?

Many production companies and TV stations have guidelines about the ethical treatment of children on set or in journalistic interviews. In the UK, for example, Channel 4 has extensive "Working & Filming with Under 18s Guidelines" that specifically state that the overarching principle of filming children should always be "that due care must be taken over the physical and emotional welfare and the dignity of people under 18 who take part in or are otherwise involved in programs."[24] These guidelines are designed for children appearing in programs either directly in an interview or otherwise featured extensively. *Honeyland*'s observational optics, however, complicate the application of these guidelines. If the directors observe and record the *natural* day-to-day life of the Sam family, would intervention equal fraud within this documentary mode? And how, returning to Crary's definition, can the "observational" indicate more than the act of looking and instead be deployed as an engaged and intentional act?

Stefanov and Kotevska, by locating themselves in a different "time zone" from their subjects, position intervention as *unnatural* and automatically in conflict with observational cinema. Writing about *Honeyland* for the Visible Evidence Forum, Bill Nichols notes that "intervention is possible but not possible within this [observational] mode." He is referring to the bees, cattle, and other distressing factors filmed ("watched") by the filmmakers and yet not "addressed."[25] There are, however, ways in which the filmmakers could have stayed true to observation and yet intervened when children were getting hurt, both observing the action and observing ethical standards.

To comparatively assess *Honeyland*'s treatment of children, we can look at *Être et Avoir* (Nicolas Philibert, 2002), a film about primary school students in rural France that also uses different observational filmmaking protocols depending on where its subjects are filmed. The film fluctuates between direct cinema filming and interviews, allowing the audience to hear the voices of the teacher as well as of the children. When the students are filmed in company of their teacher Mr. Lopez, the camera tends to be close and intimate, reflecting the warm, sheltering nature of the little school. When they are filmed in their home environments, the camera seems to be further away, suggesting a colder atmosphere. *Être et Avoir*, while arguably not free of bias, nevertheless distinguishes itself from *Honeyland* in its preservation of a singular time.[26] In filming the teacher in interviews and hearing the filmmaker's voice asking questions from off-screen, the filmmaker – and by extension, the audience – is automatically positioned as contemporaneous with the subjects. The rural school thus is not in danger of being filmed in a time far away from the present tense. Stella Bruzzi highlights *Être et Avoir* as an example of a film that foregrounds the struggle between longing for the innocence of childhood and yet inescapable maturity in the coming-of-age.[27] This type of temporal complexity cannot be found in *Honeyland*, perhaps because the children appear to only exist in the present tense.

If the filmmakers were to intervene in scenes involving the parents, they would have to acknowledge that a return to Hatidže's timeless past is not a feasible alternative to the present. The struggles the Sam family faces are not documented to posit correctives for a better future or to deal with the complicated issues related to innocence and coming-of-age; they are recorded as a symbol of the present's destruction of the past, a negative force destroying Hatidže's way of life. Fatimah Tobing Rony has written extensively about anthropology's temporal orientations. Leaning on Johannes Fabian's work, she writes that "the people studied are timeless," and "the anthropologist [is a] hidden observer, akin to the natural historian in that he or she stands at the peephole into the distant past."[28] By intervening, *Honeyland*'s filmmakers would have had to leave their "peephole" to acknowledge their own contemporaneity with the Sams and the fact that they, too, are aligned with the destructive forces of modernity according to the temporal divisions set up by their own film. In ignoring the children's pain, however, the filmmakers obscure their own temporal position. Beyond physical protection, the very act of intervention could have acknowledged the filmmakers as inhabiting the same contemporaneity as the children, and as such, having a responsibility towards their wellbeing (at the very least) on set.

The film suggests that the future of the eponymous land depends only on whether or not Hatidže can survive through her old ways, not on the interventions of a new generation.[29] According to *Honeyland*, the only way forward is the way back to Hatidže and her traditional way of life; contrary to the saying, the children are not the future in this film. The filmmakers show no interest in generational or longitudinal explorations; this would demand their acknowledgment that this rural area is not a time capsule. Scenes in which the Sams' son is drawn towards Hatidže perhaps offer a glimmer of hope that future generations could continue in Hatidže's beekeeping tradition. However, once the Sams leave, the film does not follow the boy, and we are again left with only Hatidže as a protector of the "old ways." Perhaps, if the children were given actual status as equal documentary subjects, their growth and development would clash with the idea that Hatidže's world is frozen in time or can be restored.

In their directors' statement, Stefanov and Kotevska stress over and over again how Hatidže needs to be read as one with nature, as if nature existed in a different time than mankind.[30] This denial of coevalness places the filmmakers into an anthropological tradition that Buñuel already critiqued in the 1930s, when he made *Land without Bread* (1933) to sarcastically call attention to the Spanish government's neglect of the Las Hurdes region of Spain, delivering an intentionally cruel depiction of the region's inhabitants that aimed to ironically parallel the carelessness of the government. The cultural taxonomy of observation, in this tradition, will always create different times for maker and subject for, as Ernst Bloch observed, "cultural gardens lie behind the walls of relativism. The anthropologist may watch them grow and change but whatever happens behind the walls occurs in a Time other than his."[31] The Sams, while indexing the threats of the present tense, nevertheless also seem to be encoded as living in yet a different, slightly "backward" present tense that separates them from the filmmakers. Indeed, it comes as a moment of surprise when we see one of the daughters wearing a *Frozen* T-shirt towards the end of the film, an indexical hiccup by which the visuals, for once in this film, betray the ideological construction of the Sams' temporal otherness from the filmmakers. A repression of coevalness is not an accidental byproduct; it is a political decision. Under the guise of observational film codes and temporal distance, Stefanov and Kotevska do not intervene, thereby implicating us, the audience, in the same carelessness and neglect towards these children.

Outside of the image, we have to wonder if the Sams' children received any more care than what we see in the finalized documentary. Since children are not capable of giving fully informed consent[32] and the Sams are cast as dubious figures who threaten their children and choose money over morals, one could question how the filmmakers did indeed obtain consent to film these children. After filming concluded, the Sams received a new house, a new truck, and repairs on their chimney, all paid for by the filmmakers.[33] However, as Calvin Pryluck writes, "The ethical status of responsible consent becomes obscure where what is being agreed to is only marginally for the benefit of the minor child."[34] By what we can judge from the film text, it is not clear that the children benefited from the presence of the film crew. Yes, they have a new house and a truck, but the long-term consequences of these images depicting their suffering are unknown. While watching, we might not consider how these images eventually affect the children's status in their schools, chances of getting a job, or of being ridiculed by other children. A *New York Times* article from August 2020 discusses the afterlife of *Honeyland* and its directors' continuing relationship with the protagonists. The writer brings up the dilemma, "As observers, should [filmmakers] ever help their subjects? And, as humans, how could they ever not"?[35] Yet "help" in this case seems to be restricted to the financial.

By and large, we learn about the filmmakers' relationships with their subjects after the fact, through articles and news sources that exist outside of the profilmic. In the case of *Honeyland*, the temporality that exists around the film – that is, the filmmakers' continued relationship with Hatidže and the Sams – is invisible and inconceivable within the film itself, as the temporal differences are set up to create a time within which the filmmakers could not interact with the Sams or Hatidže. Within the film, at least, observation becomes a cover for the filmmakers' neglect.

Conclusion

Throughout the film, the timelessness of nature is used to remove "unnatural" concepts such as responsibility and ethics from the discussion. By the end of the documentary, the Sams have moved on, and – even though the documentary insists that Hatidže is the only joyful presence in the children's lives – their leaving does not seem to have any visible effect on Hatidže herself. In picturesque panoramic wide-angle shots that show her still climbing mountains as she did in the beginning, she is alone with her dog, but she is not visibly mourning the departure of the Sams' children. Neither does the film. By the end, it seems as if the children were never there. Although they, like the animals, were an important part of the scenery in the film, they are still too human, too modern to fit the film's romantic vision of nature. As both natural and unnatural, their liminality makes them ultimately irrelevant to the film's dualistic ideology.

With the Sams' departure, we are left with natural timeless beauty, but an observational documentary needs more than visual beauty; it requires a strong commitment to an ethical aspiration with, among, and towards its participants. The temporality of a beautiful landscape seems to be timeless. *Honeyland*, however, is a documentary about people, not just a landscape. As such, beauty cannot be positioned as something that exists outside of time. Perhaps the real beauty of observational cinema lies in simultaneity, as shared time between filmmakers and subjects.

Notes

1 A.O. Scott, "'Honeyland' Review: The Sting and the Sweetness," *New York Times*, July 25, 2019.
2 David Sims, "A Rare Nature Documentary That Tells a Deeply Human Story," *The Atlantic*, July 25, 2019.
3 David Ehrlich, "'Honeyland' Review: Macedonia's Last Beekeeper is the Heart of Harrowing Doc about Environmental Balance," *IndieWire*, accessed March 30, 2021, https://www.indiewire.com/2019/03/honeyland-review-macedonia-beekeeper-documentary-1202054212/.
4 Michael O'Sullivan, "In 'Honeyland,' a Solitary Beekeeper Delivers a Warning about Life in and Out of Balance," *Washington Post*, August 5, 2019.
5 Louise Spence and Vinicius Navarro, *Crafting Truth: Documentary Form and Meaning* (Rutgers, NJ: Rutgers University Press, 2010), 221.
6 My approach to temporality in *Honeyland* builds on theorizations of time and the other as discussed in Johannes Fabian, *Time and the Other: How Anthropology Makes Its Object* (New York: Columbia University Press, 1983), Fatimah Tobing Rony, *The Third Eye: Race, Cinema, and Ethnographic Spectacle* (Durham, NC: Duke University Press, 1996), and Bliss Cua Lim, *Translating Time: Cinema, the Fantastic, and Temporal Critique* (Durham, NC: Duke University Press, 2009).
7 "Director's Statement," *Honeyland* Website, accessed October 26, 2020, https://honeyland.earth/story/

8 Brian Winston, "The Tradition of the Victim in the Griersonian Documentary," in *New Challenges for Documentary*, ed. Alan Rosenthal (Berkeley: University of California Press, 1988), 276.
9 As has been suggested by Dina Iordanova in "Underdevelopment, Coated in Honey," *Docalogue*, June 1, 2020, accessed March 27, 2021, https://docalogue.com/honeyland/.
10 Anna Grimshaw and Amanda Ravetz, *Observational Cinema: Anthropology, Film, and the Exploration of Social Life* (Bloomington: Indiana University Press, 2009), 7.
11 See, for example, Rony, *The Third Eye*.
12 Jay Ruby, *Picturing Culture: Explorations of Film and Anthropology* (Chicago, IL: University of Chicago Press, 2000), 8.
13 Bill Nichols, *Introduction to Documentary* (Bloomington: Indiana University Press, 2001), 109–10.
14 Quoted in Grimshaw and Ravetz, *Observational Cinema*, 6.
15 For more information on MacDougall's approach see David MacDougall, "The Visual in Anthropology" in *Rethinking Visual Anthropology*, eds. Marcus Banks and Howard Morphy (New Haven, CT: Yale University Press, 1999).
16 Nichols, *Introduction to Documentary*, 112–13.
17 Spence and Navarro, *Crafting Truth*, 195.
18 Stephen Mamber, *Cinema Verité in America: Studies in Uncontrolled Documentary* (Cambridge, MA: MIT Press, 1974), 2. Mamber uses the term "real people" to distinguish documentary subjects from professional actors.
19 Grimshaw and Ravetz, *Observational Cinema*, 8.
20 Jean Bentley, "How 'Honeyland' Documentary Found Its Beekeeping Protagonist in Rural Macedonia," *Indiewire*, September 30, 2019, https://www.indiewire.com/2019/09/honeyland-documentary-Hatidže-muratova-ida-1202176686/.
21 Jonathan Crary, *Techniques of the Observer: On Vision and Modernity in the Nineteenth Century* (Cambridge, MA: MIT Press, 1993), 5–6.
22 Grimshaw and Ravetz, *Observational Cinema*, 10–11.
23 UNICEF, "Violence against Children in East Asia and the Pacific: A Regional Review and Synthesis of Findings," UNICEF.org, accessed March 26, 2021, https://www.unicef.org/eap/media/2901/file/violence.pdf.
24 "Working & Filming with Under 18's Guidelines," Channel 4, accessed October 27, 2020, https://www.channel4.com/producers-handbook/c4-guidelines/working-and-filming-with-under-18s-guidelines.
25 Bill Nichols, "Thoughts on Raw Footage, Observational Documentaries, and the Cinema," *Visible Evidence Forum*, September 2020, https://www.visibleevidence.org/article/thoughts-on-raw-footage-observational-documentaries-and-the-cinema/.
26 Shari Kizirian, "*Être et Avoir*: The Medium and the Moment," *Senses of Cinema*, no. 60, October 2011.
27 Stella Bruzzi, "From Innocence to Experience: The Representation of Children in Four Documentary Films" in *Studies in Documentary Film* 12, no. 3 (2018): 217.
28 Rony, *The Third Eye*, 30 and 102.

29 For further writing on children as symbolic representatives of childhood, see *Symbolic Childhood*, ed. Daniel Thomas Cook (New York: Peter Lang, 2002).
30 "Director's Statement," *Honeyland* Website, accessed October 26, 2020, https://honeyland.earth/story/.
31 Ernst Bloch, quoted in Johannes Fabian, *Time and the Other: How Anthropology Makes Its Object* (New York: Columbia University Press, 2002), 52.
32 Calvin Pryluck, "Ultimately We Are All Outsiders," in *New Challenges for Documentary*, ed. Alan Rosenthal (Berkeley: University of California Press, 1988), 201.
33 Patrick Kingsley, "Film Crew Spent 3 Years in Remote Balkan Hamlet. Will They Ever Leave?" *New York Times*, August 29, 2020, https://www.nytimes.com/2020/08/29/world/europe/honeyland-north-macedonia-bees.html.
34 Pryluck, "Ultimately We Are All Outsiders," 201.
35 Kingsley, "Film Crew Spent 3 Years in Remote Balkan Hamlet. Will They Ever Leave?".

5 Feeling *a* life

Sympoietic aesthetics in *Honeyland*

Maja Manojlovic

> *A* life is everywhere, in all the moments that a given living subject goes through and that are measured by given lived objects: an immanent life carrying with it the events or singularities that are merely actualized as subjects and objects. This indefinite life does not itself have moments, close as they may be to one another, but only between-times, between moments; it doesn't just come about or come after but offers the immensity of an empty time where one sees the event yet to come and already happened, in the absolute of an immediate consciousness.[1]
>
> Gilles Deleuze

> Sympoiesis is a simple word; it means "making-with." (…) It's a word for worlding-with, in company.[2]
>
> Donna Haraway

> The cinema must film, not the world, but belief in this world, our only link.[3]
>
> Gilles Deleuze

This essay reflects on the North Macedonian film *Honeyland* (*Медена земја*, 2019), which blends observational style documentary with a narrative crafted as a "lyrical environmental fable."[4] This mélange of the incongruent documentary and fictional approaches not only generates tensions in its interpretations but also comes with scads of unanswered questions.[5] Amidst the overwhelming accolades and awards (three at the Sundance Film Festival,[6] and two Academy Award nominations for Best Documentary Feature and Best International Feature Film). *Honeyland* also

DOI: 10.4324/9781003124573-6

confronted fierce criticism for exploiting the poverty[7] of its subjects and lacking social context.[8] While this essay references the various domains of critical discourses in the media, along with interviews with the filmmakers and protagonist Hatidže, as well as audience reactions, my goal is not to assess this film's merits and flaws. Rather, I aim to articulate a confluence of *Honeyland*'s elements that, combined, create a transcendent effect of a momentary *encounter* with what Gilles Deleuze conceptualizes as "*a* life" as "pure immanence."[9]

Indeed, the alchemy of this film's aesthetic, cultural, and environmental elements responds to Deleuze's paradoxical call for cinema to film "not the world, but belief in this world,"[10] which is to film the unfilmable. I suggest that *Honeyland*, even if unwittingly, succeeds in filming the unfilmable. It does so by cinematically bringing forth a primal life-force, a "germ of life" deeply buried in the bodies of every-one and every-thing inhabiting the film: the protagonist Hatidže, her mother Nazife, their nomadic neighbors, animals, bees, and both natural and urban environments. The primal life-force reverberating through *Honeyland*'s images thus brings forth a *feeling* of being alive-in-the-world, as a *sympoietic* process of "making-with," "worlding-with, in company of" others and other things.[11] In this sense, the sympoietics of *Honeyland*'s world convey the unfilmable entanglements and inter-relationships between human and non-human beings, organic and non-organic matter, that go beyond the representational and linguistic logic of its narrative. Instead, these sympoietic entanglements of its film-world shine through the affect and gestures of every-one and every-thing in the film; they are cinematographically expressed in the textured interplay of illumination from bright sunlight to deep nocturnal darkness. Moreover, the sympoietic entanglements are expressed through the rhythms and cadences of human languages and the non-human sounds of the mountains and urbanized environments. *Honeyland* therefore offers a novel approach to making sense of our life-world[12] by way of sympoietic aesthetics.

Thinking through how to "stay with the trouble" of the global environmental crisis, Donna Haraway uses the term *sympoiesis* to propose an alternative to understanding and relating with our world. She describes sympoiesis as "collectively-producing systems that do not have self-defined spatial or temporal boundaries. Information and control are distributed among components. The systems are evolutionary and have potential for surprising change."[13] Sympoietics therefore account for the unpredictability of such evolutionary systems, in that they don't simply reproduce the known, but instead have the potential to surprise with novel assemblages that are not immediately recognizable. We could therefore understand sympoiesis as a process that produces *difference*. The latter might be described as something that is not self-evident, thus eluding our immediate recognition and understanding. Outside of general principles of representations of thought and representational practices of cinema, difference, as it figures in *Honeyland*, pertains to the domain of the *aesthetic regime of art*. According to Jacques Rancière:

> In the aesthetic regime, artistic phenomena are identified by their adherence to a specific regime of the sensible, which is extricated from its ordinary connections and is inhabited by a heterogeneous power, the power of a form of thought that has become foreign to itself...The aesthetic state is a pure instance of suspension, a moment when form is experienced for itself. Moreover, it's the moment of the formation and education of a specific type of humanity.[14]

Rancière's notion of the aesthetic regime of art therefore articulates the production of difference as it happens in artistic experimentation, bringing forth novel ideas and expressions. In this sense, Rancière's notion of the aesthetic is akin to that of Gilles Deleuze, when he suggests that aesthetic experiments[15] allow us to not only discover a different perspective on something that is already there but also to invent and think about things that have not been thought of before. Furthermore, for Rancière, aesthetics broadly refers to the *distribution of the sensible*, which

> refers to the implicit law governing the sensible order that parcels out places and forms of participation in a common world by first establishing the modes of perception within which these are inscribed. The distribution of the sensible thus produces a system of self-evident facts of perception based on the set horizons of what is visible and audible, as well as what can be said, thought, made or done.[16]

Sympoietic aesthetics can therefore be defined as bringing forth novel modes of perception. These challenge us to see, hear, say, think, make, and do things that are outside of the implied, self-evident norms. As such, sympoietic modes of perception are inhabited by a heterogeneous power, generating different kinds of the "distribution of the sensible." The latter allow for novel approaches to both perceiving and understanding complex interrelations, which ultimately reconfigure the notion of humanity in ways that foreground the role of feeling.

By working through the elements of *Honeyland*'s sympoietic aesthetics, I thus aim to unpack the experience that activates the gut-level feeling, and affect, that enable us to both think through and stay "with the trouble."[17] To that end, I build on Antonio Damasio's assertion that "due to the machinery of affect that evolution has built around feelings (…) it is not possible to talk about thinking, intelligence, and creativity in any meaningful way without factoring in feelings. (…) Feelings are *for* life regulation," and "inform us about risks," as well as "opportunities."[18] I complement Damasio's insistence on the vitality of feelings for our subjective and cultural development with Anna Gibbs's concept of "affective scripts."[19]Akin to Damasio's description of feelings, "affective scripts" operate largely outside of awareness, but "form an experiential matrix" for our "responses to and constructions of the world."[20] More importantly for my discussion of the processes involved in sympoietic aesthetics, Gibbs suggests that "in producing difference by means of cross-modal translation, affect *organizes*, both intra- and inter-corporeally, though it does so in very different ways in different cultures."[21] Such organization of affect between (intra-) and within (inter-) bodies is pivotal to the sympoietic aesthetics in *Honeyland*, and specifically to the ways in which different species and entities connect, sensorily translate, and co-create their world through the inter-corporeal extension of their intra-corporeal feelings and affects.

Honeyland's narrative revolves around Hatidže Muratova's care for her mother, as well as her sustainable methods of wild beekeeping, which honor the life of bees and are attuned to the natural environment. "Half for me, half for you," she says to the bees every time she takes their honey. The Sam family settles into the village every spring; while keeping her company, they also interfere with her environmental ethics. Their conflict culminates when Hussein mistreats his bees, taking too much honey for them to be able to survive. This happens because he accepts an offer from the entrepreneurial Safet, requiring him to produce more honey than his bees are able to make. Hussein's bees attack and kill Hatidže's entire beehive – the main source of her livelihood – causing her immense trouble.

Like filmmaking, the sympoietics of "staying with the trouble" implies a (temporally and spatially) dynamic process, involving many diverse participants, such as species, elements, landscapes, and their multiple ways of relating, "making-" and "worlding-with" each other. Interviews with filmmakers Tamara Kotevska and Ljubomir Stefanov reveal how it took their small crew (including cinematographers Fejmi Daut and Samir Ljuma) a hundred shooting days, interspersed across three years, to complete the production process. They weathered the harsh seasons in the company of Hatidže, her partly blind mother Nazife, their dog Jackie, their two cats, kittens, and the bees. The crew and Hatidže were then joined by the nomadic Sam family: Ljutvie and Hussein, with their eight children, a truck, trailer, and farm animals: cows, calves, chickens, dogs. Throughout the film, human and animal bodies and lives appear closely intertwined with one another. For example, in one of the first scenes after their arrival, we see boys rambunctiously playing in the trailer, sharing its space with several hens and their chickens. Similarly, Hatidže is surrounded with bees, her cat cuddles in her lap, and Jackie faithfully follows her on her trails.

While they communicate with their subjects in Macedonian, the filmmakers don't understand the Turkish dialect in which their subjects speak.[22] The filmmakers' decisions about what to shoot and how to edit the film were based on, as they put it, their own (intra-corporeal) "instinctual understanding" of their subjects' (inter-corporeally extended) body language: "So much of what we can see of each other is done without talking."[23] By focusing on the nonverbal expressiveness of characters' bodily movements instead of following the thread of the linguistically conveyed meaning, the camera's mechanical eye dilates. Its expanded field of vision now includes the bodies of animals, insects, plants, and things populating the film as both significant and signifying, thus dislocating the anthropocentric and verbal representational hierarchies and allowing for an aesthetic (re)distribution of the sensible. Moreover, this open view accentuates the physicality of human and non-human bodies, which culminates in the scenes where the Sam family corrals their cattle. Here, the collision of human and animal bodies' kinesthetic forces and affective intensities foregrounds their unbridled vitality. The viscerally overwhelming physicality of their bodies and the fierceness of their vitality are impossible to control and organize within the Western frameworks of representational thinking. The latter is best defined in Deleuze's *Difference and Repetition*, where he thinks through how representational thought occludes the potential for difference to emerge: "Difference becomes an object of representation always in relation to a conceived identity, a judged analogy, an imagined opposition or a perceived similitude."[24] In other words, difference, along with culturally textured nuances of interpretive potentials, gets buried in the pre-fabricated frameworks of reference. The latter are applied onto the affective, cultural, and geographic landscapes that have the fuzz trimmed off the edges of their local specificity, just to fit into the familiar and established generalizations.

All this is to say that the sympoietics of this film are not only intrinsic to its cinematic expressiveness but are also ingrained in the entirety of the filmmaking process. This is especially so when we understand this process in light of the multiple ways of relating, "making-" and "worlding-with" each other it requires from all its participants, including the film crew.[25] It's therefore easy to see how affects and feelings expressed as bodily gestures and postures emerge as the primary expressive force of this film's signifying as, and through, its sympoietic aesthetics of "making-with, in company." This aspect of *Honeyland*'s aesthetics crystallizes in the presence of Hatidže and her relationships of "worlding-with." The way she carries herself, the gentleness of her rugged hands moving bees off the honeycomb, and the enchanting rhythm of her voice calling "ma-at, ma-at, ma-at!" to the queen bee continue to move wildly diverse audiences. Her luminous life-force imbues all of her interactions in the film with a vitality that reverberates through the screen and extends beyond cinema.

Dwelling-with Hatidže, the land, and the hum of its languages

Our first encounter with Hatidže is from a distance, as a lone figure making her way across the plain. The extreme long shot shows a woman, dressed in a yellow shirt and brownish skirt, gingerly progressing on her path like a bee crawling along a stem. She climbs up the mountain with youthful agility and inadvertently startles a rabbit that jumps out of the bush. Acknowledging its presence, she stops momentarily to take a look at the rabbit's brisk trajectory. She then continues across a mountain ridge, the aerial shot revealing her body as a tiny dot against the vastness of the mountainous landscape. Finally, she brings us to her destination: hiding in the rocks, at vertiginous heights, is a beehive. Barehanded, she reaches into the buzzing hive and, with reverence, picks up honeycombs filled with bee-bodies that she carefully places into her basket. The close-up resounds with the low-pitched humming of the bees. She wraps the basket into a scarf and carries it on her back, trekking back across the mountain to her home. Kneeling, she lifts up the round basket and calls rhythmically to the queen bee: "Ma-at, mat, mat, mat!" The sound of her voice intertwines with the buzz of the bees. She leans toward the bee-filled basket, giving them her full attention. The medium shot frames her, steeped in the humming sound of the bees, communing and worlding-with the entirety of her surroundings: the land, trees, grass, fragrant flowers, and the relentless chorus of crickets and birds chirping as the sun's rays sparkle in the transparent wings of bees. This view dissolves into a close-up of her profile, looking into the distance, her lips curling into a soft smile (Figure 5.1).

This opening sequence establishes the film's space as a remote mountainous region and sets the tone of the film as environmentally accented. It introduces Hatidže's unique beekeeping style, which honors the bees' natural habitat, habits, and honey: "Half for me, half for you." According to the filmmakers, once they heard Hatidže utter this sentence, they knew this would be the core of their film's story. Thematically, this motto aligned with their original task, which was to make an environmental film for a nature conservation project about the river Bregalnica, which runs through the Ovče Pole region in the mountainous central part of

Figure 5.1 Medium shot of Hatidže, calling "Ma-at, mat!" to the queen bee and worlding-with the entirety of her surroundings.

the Republic of North Macedonia. *Honeyland* therefore mostly unfolds in Bekirlija, a remote, deserted village in that area. Besides Hatidže, Nazife, and their animal companions, the village is populated with a well, and dilapidated, half-ruined houses. Their stone fences are overgrown; the spatial boundaries between nature and the built environment are fuzzy. The converging temporalities of the (continuing) past, which has caused the modern exodus[26] from villages to urban environments, and the continuous emergence of local flora and fauna enveloping the village simultaneously produce timelessness of *durée* and the immediacy of *now*. The nomadic family and their animals enact just such an intertwining ebb and flow of space and time, none of them with "self-defined spatial or temporal boundaries," while pregnant with a "potential for surprising change."

Honeyland doesn't explicitly provide any geographic or cultural coordinates to orient the viewers in ways that most documentary or even fiction films would. This is potentially frustrating for any viewer unfamiliar with the region. Even those who are, like myself, familiar with it, could only identify certain landmarks and signs, such as the "SK" – meaning "Skopje" – marking cars' license plates. I was born in Skopje and lived there off and on for the first eight years of my life. I regularly traveled by train between Skopje, the capital of then Macedonia, where my parents lived at the time, and Ljubljana, the capital of Slovenia, where I lived with my maternal grandparents. This is why I feel at home in multicultural and multilingual environments, rich with various ethnic groups and languages mixing and overlapping in both public and private spaces. As a young girl, I enjoyed the freedom to switch fluidly between Slovenian, Serbian, and Macedonian. During family outings or when playing with kids on the street, I learned to pay attention to body language and its affective expressiveness. Despite being fluent in several languages, the prevalence of regional and ethnic dialects prevented me from understanding many of the spoken words.

When I lived in Skopje, I stayed with my paternal grandparents, my young parents, and my father's then-teenage sister in their minuscule two-bedroom apartment. It never occurred to me that the apartment was too small for all of us, because it was always filled with my grandmother Milka's laughter and the smells of her flavorful cooking, often mixed with the cigarette smoke of visiting friends. We lived across the street from the movie theater, Kino Vardar, that got jam-packed every weekend. Close by was the old train station with a half-broken clock. Its hands are to this day frozen at 5:17, bearing witness to the 1963 earthquake that devastated the city. Nobody in my family would (or would want to) define themselves as Third-World underprivileged Eastern Europeans, which surely is the label we would have had from the perspective of the West. I often accompanied my grandfather, Slobodan, to Bit Pazar (where Hatidže sells her honey), which is one of the largest outdoor markets. The market extends into the Old Bazaar (Stara čaršija), part of the city dating back to the 12th-century Ottoman Empire. Merchants come from all over Macedonia to sell their goods. I vividly remember my grandfather buying a kilo of delicious feta cheese. He picked it from myriads of sheep cheeses on display, each smelling and tasting slightly different, depending on what mountains the sheep grazed and the specific family recipes. Strolling through Bit Pazar, I overheard the sounds, rhythms, and cadences of various dialects of Macedonian, Serbian, Roma, Albanian, Turkish, Bulgarian, and even Greek.[27] Although not narratively emphasized, this multicultural and multilingual reality is visibly and audibly felt in *Honeyland*.

In an essay titled "Untangling Fiction and Reality in the Balkans,"[28] Daniel Petrick analyzes the film's "missed opportunity" to highlight the cultural and linguistic diversity specific to North Macedonia and the Balkans. "Once the norm," he writes, "mutual multilingualism in the Balkans is disappearing due to reduced contact between different language communities and the dominance of English and German." However, he also notes that "*Honeyland* shows us that the Balkans—as a place of convergence, mixture, language contact, and cultural richness—still exists." The first time the film does so is when Hatidže travels to Bit Pazar in Skopje. Buzzing from one seller to another, making her way through colorful rows containing everything from produce to nuts, clothing to hair colors, Hatidže comfortably switches from her Turkish dialect to Albanian and Macedonian. One of the merchants explains, with a slight tilt of the head, he is: "Albanian by father and Bosnian by mother." Responding in Albanian, Hatidže explains she is a Turk, and that all the Albanians and Turks from her village left a while ago. She doesn't remember those families, she says, because she was born in 1964. When she is choosing a hair color, she effortlessly speaks Macedonian with the next merchant. He advises her to get a "beautiful" chestnut color. After a brief conversation, he gifts her a fan for her sick mother, as a "souvenir from Skopje."

Honeyland indeed misses its opportunity to explicitly examine the full scale of the linguistic and cultural diversity in North Macedonia. Instead, the film's aesthetics foreground the country's richly textured embodied experiences generated by the mixing of languages and cultures. In this sense, hearing the sounds of the languages and their various affective intonations in *Honeyland* intimately resonated with who I am, my gut-level make up, and sensory memories. The postures, demeanors, and gestures of everyone Hatidže interacts with are all tuned to the registers of feeling and affect that are familiar to me, yet also remote, since I haven't visited in a very long time. However, I am still able to discern there is something distinct and substantially different about the embodied and affective texturing of communication styles we see in *Honeyland*. It's perhaps the thickness of the mixtures of culturally inflected grammars, along with the expanded registers of affectively modulated voices, that call for a certain fluidity and loosening of the emphasis on the entirely verbally constituted meanings. Indeed, this signals a different distribution of the sensible than the one I'm used to after living in Los Angeles for the past 25 years.

It's challenging to relate to *Honeyland*, its storytelling, and Hatidže without being put off by her dreadful poverty. We are asked to acknowledge and respect a system of values that looks similar to ours, but clearly is not. I felt physically nauseous and intellectually disturbed when seeing the material depravity of the Sams' and Hatidže's lives. There were many scenes where the visceral jolt of events was simply gut-wrenching, such as children being stung by bees, the girl almost drowning in the river, and calves apparently starved to death. In addition, I was particularly disturbed by Nazife's inflamed eye. I couldn't understand why the filmmakers didn't offer to take care of her. It turns out they did. In one of their interviews, they said they tried on many occasions to take Nazife to the doctor or at least bring her medication,[29] but she steadfastly refused. When I read that, my first thought was "typical," and I immediately understood why she refused. I could of course easily attribute her refusal to lack of education, superstition, and whatever else I can think of from my current Western perspective. Underlying her refusal to "be taken care of" is Nazife's resistance to being assigned a social place and a role to which she will not subscribe. Her values, traditions, and personal beliefs are simply not aligned with mine. To a degree, I recognize her position, but I will never fully understand her reasoning. And because of that, I resolved to resist, like she insistently does, the urge to reduce who she is, and her entire life, to one of poverty, deprivation, and suffering. Although bedridden, half-blind, and hard of hearing, she is clearly a sharp and willful woman, fully present, with her own view of the world.

The way she is attuned to and converses with her daughter demonstrate this. Her gestures and tone of voice add to her culturally specific and subjective cross-modal translations of her own embodiment as intertwined with other bodies, nature, and the elements. In one of their nighttime exchanges, where Hatidže places a lantern in a small window next to her bed, she asks her mother: "Do you see the light?" "Yes," responds Nazife, laying down. "I can see with this eye," pointing to her left eye, "but not with the other," as she gestures towards her bandaged right eye. Then Hatidže asks Nazife if she'd like to be taken out in the sun the next day. "I can't go out," she swiftly responds. "You can't take me out, I've become a tree." Nazife likens her embodied self to a tree, as she feels she has grown roots inside her house. When she sees her beloved daughter crying in despair because Hussein's bees killed her beehive, and destroyed the main source of her livelihood, she exclaims: "May god burn their livers!" This is another vivid metaphor, which reveals her expressive and spirited nature, directly linking a body organ to a natural element. Nazife's manner of speaking, while culturally colored and afforded by her Turkish dialect, is also subjectively flavored and shaped by her own affective scripts, distinct even from those of Hatidže. Not only does it denote her keen sense of observation, but also her own unique way of comprehending the world (Figure 5.2).

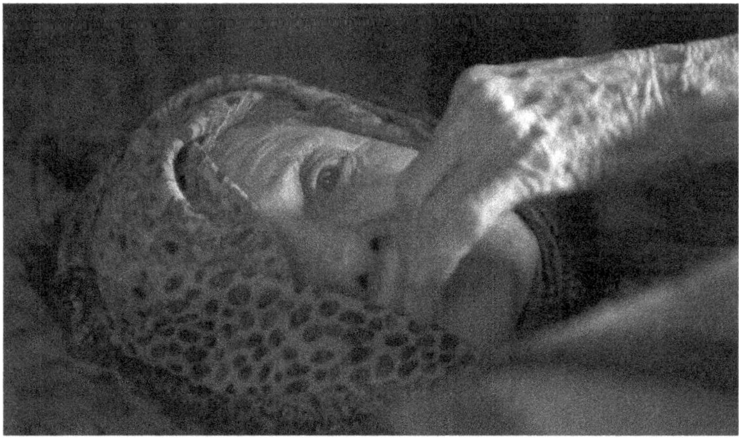

Figure 5.2 Nazife, just before she explains to her daughter that she can't take her out in the sun because she's "become a tree."

While we could reproach the filmmakers for not understanding their subjects' speech while filming, this turns out to have been a conscious decision. The film crew communicated with Hatidže and the Sams in Macedonian. However, they expressly didn't ask their subjects to speak Macedonian just so they, the filmmakers, could understand them. Instead, both Hatidže and the Sams spoke their intimate "family" language, gesturing along their own "affective scripts." Moreover, once the Sams' accepted the crew, they were a part of the family. Said Kotevska: "Once they accept you, there's no going back: You are family." However, precisely because they didn't understand their language, they had to observe differently. They did so by way of tuning in – through their own embodied feeling, affect, and instincts – to the immediacy of human interactions as they unfold in their natural environment. The filmmakers displace their focus from the meaning of words characters are saying to feeling the modulations of their voices. They follow characters' movements, tuning into their energy, speed, length, form, and direction. These visual, auditory, and kinesthetic cues, along with their alchemical reactions with the physical environment (soil, rocks, trees, birds, cats, dogs, cows) and its elements (wind, sun, rain, snow, fire), generate a feeling of deep interconnectedness and "making-with, in company."

By filming the complexity of sensory human experience without coercing it into a linguistically organized documentary rhetoric, *Honeyland* sympoietically expresses the complex entanglements of diverse and divergent affects, languages, and cultures in North Macedonia. The film's aesthetics therefore infuse its imagery and storytelling with a uniquely Macedonian "feeling" and a *poiesis* characteristic of the expressive styles of the Balkans. These are evident in the scenes unfolding in the mountainous landscapes surrounding Bekirlija, in the Bit Pazar engulfed in the urban noise of Skopje, and the fair with its Roma oil wrestling rituals, apparently happening near Durfulija, a larger, more modernized village. One way to read *Honeyland* is to allow the land to speak of a feeling, or *pathos*, generated by mixed, intertwined, and entangled cultural (hi)stories that evade a conventionally organized ethnographic description.

In the scene where the Sams first arrive at Bekirlija, we see their truck and trailer slowly crawling along the narrow, unpaved road, cutting through the greenery of the mountain valley. The rumbling of the old engine is mixed with the rhythmic chirping of crickets, singing of birds, and blowing of wind. The clacking of the metal trailer door overrides the fluttering of the wind and Hussein's voice, just before it is overpowered with the rambunctiousness of children's screams and laughter. Here, in this remote, mountainous village, we hear the overlapping of the human and non-human, as well as natural and mechanical sounds, creating a site-specific atmosphere. The gasoline exhaust mixing with fresh mountain air gives it an added charge. The smells and sounds of modern urbanism intertwine with the pre-industrial rusticism of the stone-built houses in the village. Sam's nomadic, big family-lifestyle vitally intersects with Hatidže's homebound, yet not entirely insular, living. Although different and divergent, their lives converge in the North Macedonian mountains, with their livelihoods dependent on the land of bees and their honey.

The sequence where the Sams' children and Hatidže listen to the radio gives another striking example of the entangled cultures, identities, and affects transcending both spatial boundaries and temporalities. Hussein and his sons put together a provisional antenna, and Hatidže then uses her old transistor to tune into Radio Skopje's weather news. Surrounded by Hussein, Ljutvie, three boys and a girl, she then finds a different station, which plays Joe Cocker's 1974 song "You are So Beautiful." She starts singing, mixing the melody of the British songwriter with Macedonian rhythms, with a pinch of the folk tunes native to her Turkish dialect. As the sound of the music along with the radio waves dissolve into the ether, she takes over with her singing, claps rhythmically while dancing, and cheers the kids to do the same. Her dynamic body movements, along with expressive vocalization and affective warmth, sympoietically interrelate with the radio sounds, thus transcending spatial boundaries and blurring temporalities.

The filmmakers' choice – whether intentional or not – to emplace *Honeyland*'s cinematic expressiveness within the aesthetic regime, brings forth a distribution of sounds, rhythms, movements, affects, and gestures specific to the North Macedonian experience. The film's environmentally focused narrative allows for highlighting the aesthetic that speaks to our senses, and for dimming down the representational rhetoric that categorizes according to our precepts. This demonstrates how their film emerges as a result of dwelling-with Hatidže and her life-world, rather than condescending to her and everyone else in the film.[30]

Luminosity and darkness, intimacy and immanence: a life

The filmmakers' dwelling-with Hatidže's life-world is visible and sensorily palpable in the feeling of intimacy arising from *Honeyland's* rhythms of day and night, sunlight and darkness. Characterizing the sympoietic aesthetics of worlding-with in this film, the soft force of intimacy generates unexpected distributions of the sensible. Intimate feelings and phenomena ordinarily associated with the internal externalize, while the light and dark intertwine to form "an immanent life, carrying with it the events or singularities that are merely actualized as subjects and objects."[31] This is perhaps why Petrick, after a poignant critique of the film, states, "It's unlike anything I've ever seen."[32]

It's impossible to not be moved by the opening sequence described earlier, with Hatidže calling to the queen bee in surroundings that shimmer and sparkle in the sunlight. The sunrays, momentarily refracting whenever encountering a body or a thing, engulf the entirety of the environment. The atmosphere pulsates with particles of pollen and myriads of insects, their transparent wings fluttering in the air. Vibrant with life, as revealed by natural sunlight, the air is as thick as ocean water filled with plankton particles. Sunlight, a quintessential "outside," something we look at in space and feel on the surface of our skin, here signals the depths of the internal and intimate – a feeling. Indeed, as Damasio suggests, "feelings" correspond to "*the internal state of the body within which consciousness inheres*," and "describe the inner state of life *now*."[33] In this sense, the internal embodied processes, signaling momentary feelings, interrelate and intertwine with the now-ness of the refracting sun rays outside of the body. The multitude of human and non-human, organic and non-organic, natural and mechanical bodies figuring in *Honeyland*, are enfolded in a process of worlding-with one another, as the refracting sunlight gently disperses a feeling of intimacy between all of them.

Intimacy, a feeling that dwells in enclosed spaces, thus unexpectedly spills outside, and into the brightness of daylight. Relying on natural light, Fejmi Daut and Samir Ljuma's cinematography[34] brings forth this film's richly textured atmosphere, modulated by the different kinds of intimate feelings that permeate both external and internal spaces. Light and luminosity, as well as their absence and darkness, thus both mediate and enfold the processes of worlding-with, which is imbued with various intensities and tonalities of feelings, along with life and death. As Damasio suggests, feelings pertain to a "varied quality of the state of life within a body" and can therefore be "good or bad, positive or negative, appetitive or aversive, pleasurable or painful, agreeable or disagreeable."[35] The small crew patiently dwells and worlds-with the lives of Hatidže, her mother and the Sams, as well as the entirety of their world. They film the mountains during sunrise, as its light starts dispersing the lingering shadows of the dawn. During daytime, with bright sunlight spilling over the environment, they follow Hatidže, as she quietly sets up her beehives, tenderly watching over her bees and their honeycombs and reverently collecting their honey. They are with the Sams, frantically trying to manage their cattle and bees. The degrees of sunlight, interfering with the different sizes and shades of the hills, trees, bushes, houses, fences, truck, trailer, beehives, adult and children's bodies, animals, insects, and so forth, are both shaping and mixing with the feelings generated by these bodies.

As the sunlight slowly withdraws from the land and its crevices, dusk falls, with shades dimming into shadows, and darkness swells into night. This transition is particularly clear in the sequence where Safet, a Bosnian Muslim, visits the Sams bearing a feast for Hussein's entire family. As he generously piles food onto the sunlit table, he encourages them all to "eat like kings." Ljutvie, happily surprised, asks Safet in Bosnian Serbian, "What are you doing, Safet?" to express her gratitude. Everyone is gathered around the table, bathed in the sunlight. We see children's faces happily chumming on bread, with Hussein in their midst, enjoying their happiness. As the sun goes down, and the light disperses, Hussein's face is visibly softened by the sight of his satisfied children. But as the shadows of the dusk spread across the diminished pile of food, Hussein's face darkens with worry. Safet is now talking business, the tone of his voice sharpens, as he pushes for a response from Hussein: "Do we have a deal?" he insistently demands. The shadows obscure Safet's face and amplify their affective intensity, while simultaneously reflecting the tension and ambivalence wrinkling Hussein's forehead.

The expressiveness of the different tonalities of light and dark further intensifies in the enclosed space of Hatidže's house. One of the most striking nighttime shots occurs at the beginning of the film. We first see a dark screen. A flickering light of a lantern emerges from its center and shapes into a small square window, momentarily disorienting a viewer's perception into thinking that

Hatidže is inside, looking out. Instead, the viewer is emplaced in the house's pitch-dark interior, and is looking outside. We see Hatidže as if she were a miniature. She is engulfed by darkness, holding a lantern framed by a tiny square window. She walks out of the frame and into the house, her lantern carving blots of flickering light into the darkness. She places it on the window sill and sits next to her bedridden mother. Then, the previously described conversation unfolds, where Nazife sees the light, but refuses to be taken out into the sun the next day. During their repartee, Nazife reassures her daughter she is not dying, "nor has any intention to die," because "she eats whatever she desires." Despite her apparent spunkiness, Nazife's body gestures towards death, as she refuses to leave the house or receive medication. In resonance with these ambiguous feelings, both women's faces are lit with the warm lantern light, which seems to be continuously withdrawing into the depths of nocturnal darkness.

This deep warmth of the fragile lighting, struggling to survive the darkness, gives grounding to a feeling of intimacy and intertwinement not only between the mother and daughter's lives, but also with their environment. The thickly textured, sympoietic aesthetics of this film thus invite the viewers' own feelings to interrelate and dwell-with the intimacy of the world on screen. Filled with life, this world resists judgment that would reduce its force to preassigned values and pre-made thoughts. If we dare to dive into the intimacy of dwelling-with Hatidže and worlding-with *Honeyland*, we might experience what Rancière calls the "aesthetic state." As a "pure instance of suspension, a moment when form is experienced for itself," the aesthetic state allows for the emergent difference to linger, unnamed and unformed, in the protracted moment of *a* now, oscillating between darkness and light.

"What is sometimes darkness and sometimes light, is one in nature," says Aristotle. According to Giorgio Agamben, Aristotle explains that the actuality of this nature in action is light, while "darkness is the color of potentiality."[36] By refining the texturing of light and darkness, *Honeyland*'s images make the interplay of these two aspects meaningfully palpable. Indeed, when we see darkness filling the frames of *Honeyland*'s images, we momentarily experience it as *skotos*, "the darkness that overcomes human beings at the moment of their death."[37] Allowing oneself to be engulfed by the darkness of Hatidže's and Nazife's dwelling elicits a feeling that the daily bustle of Western lifestyle seamlessly elided until the outbreak of the COVD-19 pandemic – an encounter with one's own mortality. The film's texturing of nocturnal darkness thickens with gravitas of *a* life approaching death during the winter, slowly sliding into the "immensity of an empty time," as the night echoes with the howling of the wolves and dogs. The transcendent experience of encountering *a* life, "an immanent life carrying with it the events or singularities that are merely actualized as subjects and objects" emerges from the texturing of sunlight. If, through darkness, we encounter the potential for death and mortality, the sunlight invites an encounter with the vitality of life. An encounter with *a* life in the "absolute of immediate consciousness," emerges from the interplay of the warm, golden sunlight during summertime, and the cold, crystalline whiteness of sun rays dispersing across the snow in the wintertime.

The two women are now alone in the village, with the Sam family gone for the season. The winter daylight protruding through the small square window of their house is milky cold. The whiteness of the snow glitters in the sunlight. Razor-cold air freezes Hatidže's breath as she collects the snow into worn-out tin jars. Mother and daughter dwell quietly next to one another, only to break into bickering. Nazife refuses to eat "ayran," a yogurt-based probiotic beverage and demands milk instead. Frustrated, Hatidže throws the whole plate to the floor for the cats. After what seems to be a prolonged period of silence between the two, Hatidže asks Nazife: "Do you imagine spring coming?" Her mother doesn't hear the question. After several attempts, Nazife responds with a question: "Is there spring?" "Sure there is," says Hatidže. Nazife responds after a pause: "Too many winters have passed." This was their last conversation. After that, the darkness of the winter night embraces Hatidže's mourning cries mixed with the howling of the wolves. There are so many of them that she runs out waving a torch into the dark to scare them off. "Get away, devils!," she screams with a high-pitched, animalistic voice. Hatidže dives into *skotos*, the darkness now enfolding her mother's spirit.

What naturally emerges from the darkness of Nazife's death is light. In the last minutes of the film, we are weathering the winter cold with Hatidže and her dog, Jackie, as they trek across the vast, crystal-white, snow-covered plain. This closing sequence follows a similar rhythm as the opening one. It begins with an extreme wide shot, followed by an aerial shot of the pair climbing the mountain together, where Hatidže triumphantly finds her surviving beehive tucked away in the rocks. She takes half of a honeycomb, and with much delight, shares it with Jackie who licks honey off her fingers.

The medium long shot captures their two silhouettes merging into one, just as the sunlight fades behind the mountains arching in the background. We then see Hatidže in a close-up, turning towards the white sunlight and looking out onto the horizon. Although the spring is still buried underneath a thick carpet of glittering snow, Hatidže knows it's coming. We can almost feel it sprouting, breaking through the ice-hard soil like a "germ of life," unstoppably announcing a new season. Although culminating in Hatidže's resiliency, this unruly force of life clearly comes through in the sympoietic aesthetics of this film, illuminating the entanglements of every-body and every-thing worlding-with in company of one another (Figure 5.3).

After dwelling-with Hatidže and worlding-with *Honeyland*'s sympoietically textured cinematic aesthetics, I feel alive-in-the world slightly differently than I have before. Apparently, when they first met, Hatidže told the directors she had been thinking for a long time about how to tell her story to the world.[38] The filmmakers were bold enough to experiment with storytelling that was fine-tuned to Hatidže's embodied experience of life, what is meaningful and valuable to her, and to those she is connected to. The intimate feelings brought forth by the sympoietically textured aesthetics invited me to viscerally experience that germ of life we all share. Immanent to life as such, it lives in every-body. Intolerable as our world may be, Hatidže and *Honeyland* stay with its troubles, an endeavor Donna Haraway describes as possible only: "in generative joy, terror, and

Figure 5.3 Although buried underneath the snow, we can feel the spring sprouting, breaking through the iced soil, like a "germ of life."

collective thinking."[39] By filming just that, *Honeyland* revives our belief in this world and in one another.

Notes

1 Gilles Deleuze, *Pure Immanence: Essays on A Life*, trans. Anne Boyman (New York: Zone Books, 2001), 29.
2 Donna Haraway, *Staying with the Trouble: Making Kin in Chthulucene* (Durham, NC: Duke University Press, 2016), 58.
3 Gilles Deleuze, *Cinema 2: The Time-Image*, trans. Hugh Tomlinson and Robert Galeta (Minneapolis: University of Minnesota Press, 1989), 172.
4 A.O. Scott, "'*Honeyland*' Review: The Sting and the Sweetness," *New York Times*, July 25, 2019. See also: Matthew Carey, "Oscar-Nominated '*Honeyland*' Directors on Film's Universal Themes and 'Goddess-Like Heroine," *Deadline*, February 3, 2020.
5 See, for example, Richard Brody, "'*Honeyland*,' Reviewed: A Gripping, Frustrating Documentary about a Beekeeper's Fragile Isolation," *New Yorker*, August 1, 2019.
6 Respectively, the World Cinema Grand Jury Prize for Documentary, the World Cinema Documentary Special Jury Award for Impact for Change, the World Cinema Documentary Special Jury Award for Cinematography.
7 Dina Iordanova, "Underdevelopment, Coated in Honey," *Docalogue*, June, 2020: https://docalogue.com/honeyland/. While there are others that point to the exploitation of the characters' economic conditions, Iordanova's critique effectively focuses on this aspect of the film.
8 Many critics are baffled by a lack of explanation of the social context in the film. Richard Brody's review speaks of his frustration with the film's lack of contextualization. In his own words: "(...) it's as if the reality of Muratova's life were too messy for the filmmakers' hermetic schema."
9 Deleuze, *Pure Immanence*, 29.
10 Gilles Deleuze, *Cinema 2: The Time-Image*, 172. Deleuze calls for cinema to film the belief in this world because he concedes that "this world is intolerable," which is why thought "can no longer think a world or think itself" (170). The way out of this existential impasse is: "To believe, not in a different world, but in a link between man and the world, in love or life, to believe in this as in the impossible, the unthinkable, which none the less cannot be but thought" (170). He suggests that body is the "link between man and the world." Therefore,

> We must believe in the body, but as the germ of life, the seed which splits open the paving-stones, which has been preserved and lives on in the holy shroud or the mummy's bandages, and which bears witness to life, in this world as it is.
>
> (173)

Understanding the body "the germ of life" also illuminates the notion of *a* life as immanence; an indefinite life, which becomes empirically palpable in the moment "when individual life confronts universal death." However, life should not be defined by this "single moment," because "*A* life is everywhere, in all the moments (…)" (see Deleuze, *Pure Immanence*, 28–29).

11 Haraway, *Staying with the Trouble*, 58.

12 In the phenomenological sense, life-world or *Lebenswelt* indicates: "The world as immediately and directly experienced in the subjectivity of everyday life," including "individual, social, perceptual, and practical experiences." Britannica, T. Editors of Encyclopedia. "Life-world." *Encyclopedia Britannica*, March 17, 2016.

13 Haraway further describes sympoiesis as a "word proper to complex, dynamic, responsive, situated, historical systems" (89). She attributes the term to M. Beth Dempster (92).

14 Jacques Rancière, *The Politics of Aesthetics*, trans. Gabriel Rockhill (London, New York: Continuum, 2004), 22–24.

15 Gilles Deleuze, *The Logic of Sense*, trans. Mark Lester with Charles Stivale, ed. Constantin V. Boundas (New York: Columbia University Press, 1990), 260.

16 Rancière, 82, 85.

17 Haraway, 92.

18 Antonio Damasio, *The Strange Order of Things: Life, Feeling, and the Making of Cultures* (New York: Pantheon Books, 2018), 139.

19 Anna Gibbs, "After Affect: Sympathy, Synchrony, and Mimetic Communication," in *The Affect Theory Reader*, eds. Melissa Gregg and Gregory J. Seigworth (Durham, NC and London: Duke University Press, 2000), 195.

20 Gibbs, 195.

21 Gibbs, 196.

22 Notably, they needed four translators to properly translate the characters' conversations. As Kotevska explained in an interview with Christy Lemire, the ancient dialect of the Quizilbashi tribe Hatidže and her mother were speaking was difficult to decipher even for the Turkish translators. In the same interview, Stefanov also comments on the difficulty contemporary Turkish speakers have with understanding this dialect. Interview with Christy Lemire, Film Critic, KPCC's *Film Week. Viva Videography*, February 20, 2020 (9:15–9:35): https://www.youtube.com/watch?v=1z50i0dwA-w.

23 Daniel Eagan, "'It's a Choice: To Be a Filmmaker or a Human:' Tamara Kotevska and Ljubomir Stefanov on *Honeyland*," *Filmmaker Magazine*, August 22, 2019.

24 Gilles Deleuze, *Difference and Repetition*, trans. Paul Patton (New York: Columbia University Press, 1994), 138. Deleuze's critique of representational thinking as stifling the notion of "difference in itself" resonates with Rancière's move from representational regime to the aesthetic regime of the arts, where the latter allows for the difference to emerge.

25 The filmmakers continue to be involved in the lives of Hatidže and the Sam family. They used their awards to purchase a house for Hatidže, a new truck and trailer for Sams. They also continue to help with school supplies for their kids. In addition to being personally involved, they arranged for a foundation so that social workers can, "help both

families overcome a never-ending list of logistical and social challenges, including setting up bank accounts and enrolling them in social security." Producer Atanas Georgiev explained: "As soon as we realized *Honeyland* would be very successful, we thought we had to do something." He felt this was "a kind of payback" to the protagonists of the film for participating in it. See Patrick Kinsley, "Film Crew Spent 3 Years in Remote Balkan Hamlet. Will they Ever Leave?" *New York Times*, August 29, 2020.

26 When speaking to a vendor in Skopje's Bit Pazar, Hatidže states that most of the Turkish families from her village left. One plausible explanation for this is that in the early 1950s Yugoslavia and Turkey agreed to open the borders for all Turkish people on the territory of Yugoslavia to repatriate to Turkey. Many people took that opportunity to reunite with their extended families. For a more detailed explanation, see Daniel Petrick, "Untangling Fiction and Reality in the Balkans," *Boston Review: A Political and Literary Forum*, December 17, 2020.

27 A combination of this diversity, mixing of languages, and co-existence of cultures and religions specific to the territory of the former Yugoslavia, is why I identified my nationality as Yugoslavian (while this option still existed, of course). In my mind, it denoted an experientially pluralist national identity, allowing for my multicultural and multilingual (way of) becoming.

28 Petrick, "Untangling Fiction and Reality in the Balkans."

29 Daniel Eagan asked the directors whether they interfered with the lives of their subjects. They responded they didn't, except offering medical care for Nazife, who refused. Daniel Eagan, "'It's a Choice: To be a Filmmaker or a Human:' Tamara Kotevska and Ljubomir Stefanov on *Honeyland*," *Filmmaker Magazine*, August 22, 2019.

30 I didn't see *Honeyland*'s narrative as imposing a straightforward protagonist/antagonist in terms of Hatidže being "good" and the Sam family "the villains." Indeed, the filmmakers have explained in interviews that they had no intention of vilifying Hussein for being determined to take care of his family.

31 Deleuze, *Pure Immanence*, 29.

32 Petrick, "Untangling Fiction and Reality in the Balkans."

33 Damasio, *The Strange Order of Things*, 158 (author's emphasis).

34 Fejmi Daut explains that they filmed in a "low-cost, low-profile" approach, which allowed them to film "a story no proper film crew could have accessed, catching stunning natural scenes and light." He also notes he was barely able to "squeeze himself and the handheld gear into the cottage of the protagonist for interior shots." "There was no place to put a tripod. I was all the time behind the door. That was the only way to shoot inside the house," says Daut. Will Tizard, "'Honeyland' DP on Low-Fi Shooting with High-Powered Storytelling," *Variety*, November 25, 2019.

35 Damasio, 158.

36 Giorgio Agamben, *Potentialities: Collected Essays in Philosophy*, trans. Daniel Heller-Roazen (Stanford, CA: Stanford University Press, 1999), 180.

37 "In Homer, skotos is the darkness that overcomes human beings at the moment of their death. Human beings are capable of experiencing this skotos." Agamben, *Potentialities*, 181.

38 When asked how they decided on the star of the film, given they could have chosen her brothers, who were performing acrobatic rope climbing with their bee-keeping techniques, Kotevska responded that Hatidže was the obvious choice. Why? Because she openly expressed to the film crew that she had been wanting to share her life story for a very long time. Apparently, it was her dream that one day, TV journalists would find her, so she could tell her life story to the world. As Kotevska says (7:10–7:40):

> She is born with this great enthusiasm for life … It was interesting to see how everyone in that place is in a kind of deep sleep, even the cats … they don't move, don't react. And this woman has all the joy in the world inside herself. And this is something you don't meet anywhere, not in the big cities.

The directors spoke about this on several occasions: at *Honeyland*'s premiere at the *Frames of Representation 2019* festival in London, when interviewed by Dennis Lim, director of programming at the Film Society of Lincoln Center, as well as with Christy Lemire, Film Critic, KPCC's *Film Week*. *Viva Videography*, February 20, 2020 (2:20–3:28).

39 Haraway, 31.

Bibliography

Agamben, Giorgio. *Potentialities: Collected Essays in Philosophy*. Edited and translated by Daniel Heller-Roazen. Stanford: Stanford University Press, 1999.

Anderson, Nicole. "Pre- and Posthuman Animals: The Limits and Possibilities of Animal-Human Relations." In *Posthumous Life: Theorizing Beyond the Post-Human*, edited by Jami Weinstein and Claire Colebrook, 17–42. New York: Columbia University Press, 2017.

Badia, Lynn, Marija Cetinić, and Jeff Diamanti, eds. *Climate Realism: The Aesthetics of Weather and Atmosphere in the Anthropocene*. New York: Routledge, 2021.

Balsom, Erika. "The Reality-Based Community." *E-Flux Journal* 83 (June 2017). https://www.e-flux.com/journal/83/142332/the-reality-based-community/.

Barbash, Ilisa, and Lucien Castaing-Taylor. *The Cinema of Robert Gardner*. Oxford; New York: Berg, 2007.

Bentley, Jean. "How 'Honeyland' Documentary Found Its Beekeeping Protagonist in Rural Macedonia." *Indiewire*, September 30, 2019.

Britannica, T. Editors of Encyclopedia. "Life-world." *Encyclopedia Britannica*, March 17, 2016.

Brody, Richard. "'Honeyland,' Reviewed: A Gripping, Frustrating Documentary about a Beekeeper's Fragile Isolation." *New Yorker*, August 1, 2019.

Brown, Eric C. "Introduction." In *Insect Poetics*, edited by Eric C. Brown. Minneapolis: University of Minnesota Press, 2006.

Bruzzi, Stella. "From Innocence to Experience: The Representation of Children in Four Documentary Films." *Studies in Documentary Film* 12, no. 3 (2018): 208–24.

Buber, Martin. *I and Thou*, translated by Ronald Gregor Smith. Edinburgh: T. & T. Clark, 1937.

Buchanan, Brett. *Onto-Ethologies: The Animal Environments of Uexküll, Heidegger, Merleau-Ponty, and Deleuze*. Albany: SUNY Press, 2008.

Burgess, Diane. "Capturing Film Festival Buzz: The Methodological Dilemma of Measuring Symbolic Value." *Necsus – European Journal of Media Studies* 9, no. 2 (2020): 225–47.

Cagle, Chris. "*Kedi*: Crossover Documentary as Popular Art Cinema." In *Kedi: A Docalogue*, edited by Jaimie Baron and Kristen Fuhs, 68–85. New York: Routledge, 2021.

Cardilli, Darianna. "'Honeyland': A Mesmerizing Film on Humans, Nature and Bees." *International Documentary Association,* January 16, 2020.

Carey, Matthew. "Oscar-Nominated 'Honeyland' Directors on Film's Universal Themes and 'Goddess-Like Heroine." *Deadline*, February 3, 2020.

Chadwick, Samara. "Building Networks: An Interview with Sandra J. Ruch, Director Emeritus of the International Documentary Association." In *Documentary Film Festivals: Changes, Challenges, Professional Perspectives* (Vol. 2), edited by Aida Vallejo and Ezra Winton, 167–74. London: Palgrave Macmillan, 2020.

Channel 4. "Working & Filming with Under 18's Guidelines." https://www.channel4.com/producers-handbook/c4-guidelines/working-and-filming-with-under-18s-guidelines.

Cimatti, Felice. "From Ontology to Ethology: Uexküll and Deleuze & Guattari." In *Jakob von Uexküll and Philosophy: Life, Environments, Anthropology*, edited by Kristian Köchy and Francesca Michelina, 172–87. London: Routledge, 2019.

Clifford, James. *The Predicament of Culture: Twentieth-Century Ethnography, Literature, and Art*. Cambridge, MA: Harvard University Press, 1988.

Cook, Daniel Thomas, ed. *Symbolic Childhood*. New York: Peter Lang, 2002.

Crary, Jonathan. *Techniques of the Observer: On Vision and Modernity in the Nineteenth Century*. Cambridge, MA: MIT Press, 1993.

Damasio, Antonio. *The Strange Order of Things: Life, Feeling, and the Making of Cultures*. New York: Pantheon Books, 2018.

De Luca, Tiago, and Nuno Barradas Jorge. "From Slow Cinema to Slow Cinemas." In *Slow Cinema*, edited by Tiago de Luca and Nuno Barradas, 1–21. Edinburgh: Edinburgh University Press, 2016.

De Valck, Marijke. *Film Festivals: From European Geopolitics to Global Cinephilia*. Amsterdam: Amsterdam University Press, 2007.

———. "Fostering Art, Adding Value, Cultivating Taste: Film Festivals as Sites of Cultural Production." In *Film Festivals: History, Theory, Method, Practice*, edited by Marijke De Valck, Brendan Kredell and Skadi Loist, 100–16. London: Routledge, 2016.

De Valck, Marijke, and Mimi Soeteman. "'And the Winner Is…' What Happens Behind the Scenes of Film Festival Competitions." *International Journal of Cultural Studies* 13, no. 3 (2010): 290–307.

Del Rio, Elena. *The Grace of Destruction: A Vital Ethology of Extreme Cinemas*. New York: Bloomsbury, 2016.

Deleuze, Gilles. *Cinema 2: The Time-Image*. Translated by Hugh Tomlinson and Robert Galeta. Minneapolis: University of Minnesota Press, 1989.

———. *The Logic of Sense*. Translated by Mark Lester with Charles Stivale. Edited by Constantin V. Boundas. New York: Columbia University Press, 1990.

———. *Difference and Repetition*. Translated by Paul Patton. New York: Columbia University Press, 1994.

———. *Pure Immanence: Essays on a Life*. Translated by Anne Boyman. New York: Zone Books, 2001.

"Director's Statement." *Honeyland* Website, https://honeyland.earth/story/.

Documentary Weekly. "Interview with Honeyland Cinematographer Samir Ljuma." *YouTube*, November 15, 2019. https://www.youtube.com/watch?v=hn2VunepVtc.

Eagan, Daniel. ""It's a Choice: To Be a Filmmaker or a Human": Ljubomir Stefanov and Tamara Kotevska on *Honeyland*." *Filmmaker Magazine*, August 22, 2019.

Ehrlich, David. "'Honeyland' Review: Macedonia's Last Beekeeper Is the Heart of Harrowing Doc about Environmental Balance." *IndieWire*, March 28, 2019.

Elsaesser, Thomas. *European Cinema: Face to Face with Hollywood*. Amsterdam: Amsterdam University Press, 2005.

Fabian, Johannes. *Time and the Other: How Anthropology Makes Its Object*. New York: Columbia University Press, 1983.

"Fable, n." In *Oxford English Dictionary Online*. Oxford University Press. http://www.oed.com/view/Entry/67384.

Falicov, Tamara L. "The 'Festival Film': Film Festival Funds as Cultural Intermediaries." In *Film Festivals: History, Theory, Method, Practice*, edited by Marijke De Valck, Brendan Kredell and Skadi Loist, 209–29. London: Routledge, 2016.

Film at Lincoln Center. "Tamara Kotevska and Ljubomir Stefanov on *Honeyland*, Family & Capturing Macedonia." YouTube, April 24, 2019. https://youtu.be/_ES4mfYYsII.

France 24. "Hatidže, the Macedonian Beekeeper Charming Hollywood." February 6, 2020. https://www.france24.com/en/20200206-hatidze-the-macedonian-beekeeper-charming-hollywood.

Gaines, Jane, and Michael Renov, eds. *Collecting Visible Evidence*. Minneapolis: University of Minnesota Press, 1999.

Gaycken, Oliver. *Devices of Curiosity: Early Cinema and Popular Science*. Oxford: Oxford University Press, 2015.

Ghosh, Amitav. *The Great Derangement: Climate Change and the Unthinkable*. Chicago, IL: University of Chicago Press, 2016.

Gibbs, Anna. "After Affect: Sympathy, Synchrony, and Mimetic Communication." In *The Affect Theory Reader*, edited by Melissa Gregg and Gregory J. Seigworth, 186–205. Durham, NC and London: Duke University Press, 2000.

Ginsburg, Faye. "Decolonizing Documentary On-Screen and Off: Sensory Ethnography and the Aesthetics of Accountability." *Film Quarterly* 72, no. 1 (2018): 39–49.

Ginsburg, Faye D, Lila Abu-Lughod, and Brian Larkin, eds. *Media Worlds: Anthropology on New Terrain*. Berkeley: University of California Press, 2002.

Godmilow, Jill. "Kill the Documentary as We Know It." *Journal of Film and Video* 54, no. 2/3 (2002): 3–10.

Grimshaw, Anna, and Amanda Ravetz. *Observational Cinema: Anthropology, Film, and the Exploration of Social Life*. Bloomington: Indiana University Press, 2010.

Gruber, Jacob. "Ethnographic Salvage and the Shaping of Anthropology." *American Anthropologist* 72, no. 6 (1970): 1289–99.

Hammett-Jamart, Julia, Petar Mitric, and Eva Novrup Redvall, eds. *European Film and Television Co-Production: Policy and Practice*. London: Palgrave Macmillan, 2018.

Haraway, Donna. *Staying with the Trouble: Making Kin in Chthulucene*. Durham, NC: Duke University Press, 2016.

Hochman, Brian. *Savage Preservation: The Ethnographic Origins of Modern Media Technology*. Minneapolis: University of Minnesota Press, 2014.

Hollingsworth, Cristopher. *Poetics of the Hive: Insect Metaphor in Literature*. Iowa City: University of Iowa Press, 2001.

"Honeyland: Press Notes." *Honeyland* website. https://honeyland.earth/press/.

Hongisto, Ilona, Kaisu Hynnä-Granberg, and Annu Suvanto. "The Invention of Northeastern Europe: The Geopolitics of Programming at Documentary Film Festivals." In *Documentary Film Festivals: Changes, Challenges, Professional Perspectives* (Vol. 2), edited by Aida Vallejo and Ezra Winton, 73–91. London: Palgrave Macmillan, 2020.

ICA. "Frames of Representation 2019: Honeyland Intro + Q&A." YouTube, April 27, 2019. https://youtu.be/Gj3s7E0bE_w.

Iglesias, Eulàlia. "Documentaries at the Cannes International Film Festival: Fahrenheit 9/11 and Beyond." In *Documentary Film Festivals: Changes, Challenges, Professional Perspectives* (Vol. 2), edited by Aida Vallejo and Ezra Winton, 113–30. London: Palgrave Macmillan, 2020.

Iordanova, Dina. "The Film Festival Circuit." In *Film Festival Yearbook 1: The Festival Circuit*, edited by Dina Iordanova and Ragan Rhyne, 23–39. St Andrews: St Andrews Film Studies with College Gate Press, 2009.

———. "The Film Festival as an Industry Node." *Media Industries Journal* 1, no. 3 (2015): 7–11.

———. "Underdevelopment, Coated in Honey." *DOCALOGUE*, June 1, 2020. https://docalogue.com/honeyland/.

Ivakhiv, Adrian. "Green Film Criticism and Its Futures." *ISLE: Interdisciplinary Studies in Literature and Environment* 15, no. 2 (Summer 200): 1–28.

Juhasz, Alexandra, and Jesse Lerner, eds. *F Is for Phony: Fake Documentary and Truth's Undoing*. Minneapolis: University of Minnesota Press, 2006.

Kellert, Stephen R. "The Biological Basis for Human Values of Nature." In *The Biophilia Hypothesis*, edited by Stephen R. Kellert and O. Edward Wilson, 42–69. Washington, DC: Island Press, 1993.

———. "Values and Perceptions of Invertebrates." *Conservation Biology* 7, no. 4 (1993): 845–55.

Kingsley, Patrick. "Film Crew Spent 3 Years in Remote Balkan Hamlet. Will They Ever Leave?" *The New York Times*, August 29, 2020, sec. World.

Kizirian, Shari. "*Être et Avoir*: The Medium and the Moment." *Senses of Cinema*, Issue 60, October 2011.

Kovacevic, Tamara. "Honeyland: Life Lessons from Europe's Last Wild Beekeeper." *BBC News*, February 9, 2020, sec. Entertainment & Arts.

Kuehner, Jay. "Interviews | Keeper of Sheep Lucien Castaing-Taylor on Sweetgrass." *Cinema Scope*. https://cinema-scope.com/cinema-scope-magazine/1107/.

Kukiełko-Rogozińska, Kalina. "Following the Footprints of Edward S. Curtis: A Tale of the Vanishing Race." *Przegląd Socjologii Jakościowej* 16, no. 2 (May 31, 2020): 36–45.

Lastra, James F. "Why Is This Absurd Picture Here? Ethnology/Heterology/Buñuel." In *Rites of Realism: Essays on Corporeal Cinema*, edited by Ivone Margulies, 185–214. Durham, NC: Duke University Press, 2003.

Lee, Toby. "Beyond the Ethico-Aesthetic: Toward a Re-Valuation of the Sensory Ethnography Lab." *Visual Anthropology Review* 35, no. 2 (2019): 138–47.

Lemire, Christy. KPCC's *Film Week*. *Viva Videography*, February 20, 2020 (9:15–9:35): https://www.youtube.com/watch?v=1z50i0dwA-w.

Lim, Bliss Cua. *Translating Time: Cinema, the Fantastic, and Temporal Critique*. Durham, NC: Duke University Press, 2009.

Lippit, Akira Mizuta. *Electric Animal: Toward a Rhetoric of Wildlife*. Minneapolis: University of Minnesota Press, 2000.

Loist, Skadi. "The Film Festival Circuit. Networks, Hierarchies, and Circulation." In *Film Festivals: History, Theory, Method, Practice*, edited by Marijke De Valck, Brendan Kredell and Skadi Loist, 49–64. London: Routledge, 2016.

MacDonald, Scott. "Ruminating on *Sweetgrass*: An Interview with Ilisa Barbash and Lucien Castaing-Taylor." *Framework: The Journal of Cinema and Media*, 2012. https://www.frameworknow.com/news/interview-with-ilisa-barbash-and-lucien-castaing-taylo.

———. *American Ethnographic Film and Personal Documentary: The Cambridge Turn*. Berkeley: University of California Press, 2013.

MacDougall, David. "The Visual in Anthropology." In *Rethinking Visual Anthropology*, edited by Marcus Banks and Howard Morphy, 276–95. New Haven, CT: Yale University Press, 1999.

———. "Beyond Observational Cinema." In *Principles of Visual Anthropology*, edited by Paul Hockings, Third, 115–32. New York: Walter de Gruyter, 2003.

———. *The Corporeal Image: Film, Ethnography, and the Senses*. Princeton, NJ: Princeton University Press, 2006.

———. "Observational and Participatory Approaches. Colin Young, Ethnographic Film and the Film Culture of the 1960s." In *Memories of the Origins of Ethnographic Film*, edited by Beate Engelbrecht, 123–31. New York: Peter Lang, 2007.

MacDougall, David, and Lucien Castaing-Taylor. *Transcultural Cinema*. Princeton, NJ: Princeton University Press, 1998.

Malitsky, Joshua. "Science and Documentary: Unity, Indexicality, Reality." *Journal of Visual Culture* 11, no. 3 (2012): 237–57.

Mamber, Stephen. *Cinema Verité in America: Studies in Uncontrolled Documentary*. Cambridge, MA: MIT Press, 1974.

Marks, Laura U. *The Skin of the Film: Intercultural Cinema, Embodiment, and the Senses*. Durham, NC: Duke University Press, 2000.

Mead, Margaret. "Visual Anthropology in a Discipline of Words." In *Principles of Visual Anthropology*, edited by Paul Hockings, 3–12. New York: Walter de Gruyter, 2003.

Menand, Louis. *The Free World: Art and Thought in the Cold War*. New York: Farrar, Straus and Giroux, 2021.

Mullarkey, John. "Animal Spirits: Philosomorphism and the Background Revolts of Cinema." *Angelaki: Journal of the Theoretical Humanities* 18, no. 1 (2013): 11–29.

Nichols, Bill. *Representing Reality Issues and Concepts in Documentary*. Bloomington: Indiana University Press, 1991.

———. "Thoughts on Raw Footage, Observational Documentaries, and the Cinema." *Visible Evidence Forum*, September 2020. https://www.visibleevidence.org/article/thoughts-on-raw-footage-observational-documentaries-and-the-cinema/.

———. *Introduction to Documentary*. Bloomington: Indiana University Press, 2001.

O'Connell, Jennifer M. J., and Annelies van Noortwijk. "Selecting Films for Festivals and Documentary Funds: An Interview with Independent Film Programmer and Advisor Rada Šešić." In *Documentary Film Festivals: Changes, Challenges, Professional Perspectives* (Vol. 2), edited by Aida Vallejo and Ezra Winton, 175–84. London: Palgrave Macmillan, 2020.

O'Falt, Chris. "Light, Cinema, Conflict: The Deceptively Simple Brilliance of *Honeyland*." *IndieWire*, January 6, 2020.

O'Sullivan, Michael. "'Honeyland' Review: A Documentary about a Macedonian Beekeeper Delivers a Message about Life out of Balance." *The Washington Post*, August 5, 2019.

Ozturk, Mustafa. "Documentary Carries Macedonian Beekeeper Out of Poverty." *Anadolu Agency*, January 23, 2020.

Peranson, Mark. "First You Get the Power, Then You Get the Money: Two Models of Film Festivals." *Cineaste* 33, no. 3 (2008): 37–43.

Petrick, Daniel. "Untangling Fiction and Reality in the Balkans." *Boston Review*, December 17 2020.

Pryluck, Calvin. "Ultimately We Are All Outsiders: The Ethics of Documentary Filmmaking." In *New Challenges for Documentary*, edited by Alan Rosenthal, 194–208. Berkeley: University of California Press, 1988.

Rancière, Jacques. *The Politics of Aesthetics*. Translated by Gabriel Rockhill. London; New York: Continuum, 2004.

Renov, Michael. *Theorizing Documentary*. New York; London: Routledge, 1993.

———, ed. *The Subject of Documentary*. Minneapolis: University of Minnesota Press, 2004.

Rice, Andy. "Being There Again: Reenacting Camerawork in 'In Country' (2014)." *JumpCut: A Review of Contemporary Media* 58 (Spring 2018). https://www.ejumpcut.org/archive/jc58.2018/RiceVietnamEnactment/index.html

Robinson, Gene E. "Darwinian Bee-Keeping: Lessons from the Wild." *Nature* 571 (2019): 34–36.

Rony, Fatimah Tobing. *The Third Eye: Race, Cinema, and Ethnographic Spectacle*. Durham, NC: Duke University Press, 2004.

Ruby, Jay. "Introduction: A Reevaluation of Robert J. Flaherty, Photographer and Filmmaker." *Studies in Visual Communication* 6, no. 2 (June 1980): 2–4.

———. *Picturing Culture: Explorations of Film and Anthropology*. Chicago, IL: University of Chicago Press, 2000.

Sarris, Andrew. "Catch as Catch Cannes: The Moles and the Moths." *The Village Voice*, December 6, 1978, 39–40.

Schultz-Figueroa, Benjamin. "From Cat to Clowder: *Kedi* in the Anthropocene." In *Kedi: A Docalogue*, edited by Jaimie Baron and Kristen Fuhs, 4–18. New York Routledge, 2021.

Scott, A.O. "'Honeyland' Review: The Sting and the Sweetness." *The New York Times*, July 25, 2019.

Scott, A.O. and Manohla Dargis. "Best Films of 2019." *The New York Times*, December 4, 2019.

Seeley, Thomas D. *The Lives of Bees: The Untold Story of the Honey Bee in the Wild*. Princeton, NJ: Princeton University Press, 2019.

Shaviro, Steven. "Slow Cinema vs Fast Films." *The Pinocchio Theory* (blog), May 12, 2010. http://www.shaviro.com/Blog/?p=891.

Sims, David. "A Rare Nature Documentary That Tells a Deeply Human Story." *The Atlantic*, July 25, 2019.

Sleigh, Charlotte. "Inside Out: The Unsettling Nature of Insects." In *Insect Poetics*, edited by Eric C. Brown, 281–97. Minneapolis: University of Minnesota Press, 2006.

Smaill, Belinda. "Documentary Film and Animal Modernity in *Raw Herring* and *Sweetgrass*." *Australian Humanities Review* 57 (2014): 61–80.

Smith, Ronald Gregor. "Translator's Introduction." In Martin Buber, *I and Thou*, v–xiii. Edinburgh: T. & T. Clark, 1937.

Spence, Louise and Vinicius Navarro. *Crafting Truth: Documentary Form and Meaning*. Rutgers, NJ: Rutgers University Press, 2010.

Stepeni, 360. "Finding the Star of Honeyland." YouTube, January 23, 2020. https://youtu.be/dj9mT91fe64.

Stevenson, Lisa, and Eduardo Kohn. "'Leviathan': An Ethnographic Dream." *Visual Anthropology Review* 31, no. 1 (2015): 49–53.

"Story – HONEYLAND." https://honeyland.earth/story/.

Swiss Agency for Development and Cooperation. "Press Release: Honeyland Oscar Nominee." January 14, 2020.

Taubin, Amy. "Swarm and Tender." *Artforum*, July 26, 2019. https://www.artforum.com/film/amy-taubin-on-honeyland-2019-80406

Tizard, Will. "*Honeyland* DP on Low-Fi Shooting with High-Powered Storytelling." *Variety*, November 25, 2019.

Turner, Fred. *From Counterculture to Cyberculture: Stewart Brand, the Whole Earth Network, and the Rise of Digital Utopianism*. Chicago, IL: University of Chicago Press, 2006.

UNICEF. "Violence against Children in East Asia and the Pacific: A Regional Review and Synthesis of Findings." https://www.unicef.org/eap/media/2901/file/violence.pdf.

United Nations' "Convention of Biological Diversity." 1992. https://www.un.org/en/observances/biological-diversity-day/convention.

Vallejo, Aida. "Industry Sections: Documentary Film Festivals between Production and Distribution." *Iluminace* 26, no. 1 (2014): 65–82.

———. "Defining Documentary in the Festival Circuit: A Conversation with Bill Nichols." In *Documentary Film Festivals: Methods, History, Practice* (Vol. 1), edited by Aida Vallejo and Ezra Winton, 19–28. London: Palgrave, 2020.

———. "Idfa's Industry Model: Fostering Global Documentary Production and Distribution." In *Documentary Film Festivals: Changes, Challenges, Professional Perspectives* (Vol. 2), edited by Aida Vallejo and Ezra Winton, 23–54. London: Palgrave, 2020.

———. "Rethinking the Canon: The Role of Film Festivals in Shaping Film History." *Studies in European Cinema* 17, no. 2 (2020): 155–69.

Vindt, Lidiya. "The Fable as Literary Genre." Translated by Miriam Gelfand and Ray Parrott. *Ulbandus Review* 5 (1987): 88–108.

Von Uexküll, Jakob. *A Foray into the Worlds of Animals and Humans: With a Theory of Meaning*. Minneapolis: University of Minnesota Press, 2010.

Walker, Janet. *Trauma Cinema: Documenting Incest and the Holocaust*. Berkeley: University of California Press, 2005.

Williams, Raymond. *The Long Revolution*. Broadview Press, 2001.

Winston, Brian. "The Tradition of the Victim in the Griersonian Documentary." In *New Challenges for Documentary*, edited by Alan Rosenthal, 269–87. Berkeley: University of California Press, 1988.

Yalom, Irvin D. *Existential Psychotherapy*. New York: Basic Books, 1980.

Young, Colin. "Observational Cinema." In *Principles of Visual Anthropology*, edited by Paul Hockings, 99–114. New York: Walter de Gruyter, 2003.

Contributor Bios

Ilona Hongisto is a Professor in Film Studies at the Norwegian University of Science and Technology (NTNU, Trondheim). She is also affiliated with Macquarie University (Sydney, Australia), University of Turku (Finland), and Aalto University (Finland). Hongisto works across Film and Media Studies, specializing in documentary cinema and its threshold with the fictitious. Focusing on questions of fabulation, speculation, and imagination, she works towards redefining the work of documentary media in the contemporary world. Her most recent work focuses on the functions of fabulation in post-1989 Eastern European documentary cinema. In conjunction with this work, she is developing her research profile in the direction of film festival and production studies. Hongisto is the author of *Soul of the Documentary: Framing, Expression, Ethics* (Amsterdam University Press, 2015), and her work has been published in journals such as *Necsus, Studies in Documentary Film, Journal of Aesthetics and Culture*, and *Cultural Studies Review*.

Linnéa J. Hussein is a Clinical Assistant Professor of Visual Cultures in Liberal Studies at New York University. Her current research project, *The Cinematic Straitjacket*, examines discourses on mental illness, race, and disability in fiction, documentary, and news media, to reframe censorship as acts of restriction that privilege comfort and protection over the right to self-represent. Before joining New York University, she was a Visiting Assistant Professor in Film Studies at Connecticut College, where she also served as a Faculty Fellow at the Ammerman Center for Arts and Technology. Her articles and reviews have appeared in *Film Quarterly, The New Inquiry, Social Text*, and *Film & History*.

Selmin Kara is an Associate Professor of Film and New Media Studies at OCAD University in Toronto. Her primary research interests are digital aesthetics and ecological sensibilities in cinema as well as the use of sound and new technologies in contemporary documentary. She is the co-editor of *Contemporary Documentary* and *Cybermedia: Science, Sound, and Vision.*

Maja Manojlovic is a Lecturer and Faculty Advisor for Writing II Pedagogy at the UCLA Writing Programs. Her background is in Cinema and Media Studies, specializing in phenomenology, aesthetics, and cultures of digital film and emerging media. She has published on digital aesthetics and 3-D cinema and presented her work at international conferences. Her current research focuses on XR aesthetics, their ethical applications as pedagogical tools for immersive writing and environmental issues, along with their use for creative self-expression. She teaches courses ranging from *Videogame Rhetoric and Design* to *XR Technologies, Immersive Environments, and Embodied Experience*. She is developing three XR experiences: *Tongva VR/AR*, supported by the UCLA Digital Research Accelerator (2019–20); *Scribe VR*, supported by the UCLA Faculty Innovation Fellows Program (2019–20); and *Re-connect! The Amazon Medicine Garden* (2018–) resulting from her participation in the 2018 Oculus Launch Pad for VR developers. She is also a co-founder and coordinator of the UCLA XR Initiative and an affiliate faculty of the UCLA Laboratory for Environmental Narrative Strategies (LENS).

Andy Rice is an Assistant Professor in Film Studies and Media and Communication in the Department of Media, Journalism & Film at Miami University in Ohio. A critical media theorist and nonfiction filmmaker, his research crosses foci on performative documentary, sensory ethnographic film, digital technologies, US social movements, and media practice pedagogy. He has published in *Jump Cut*, *The Scholar and Feminist Online*, *Journal of Film and Video*, and *Senses of Cinema*, among other venues, and also co-produced, shot, and edited the award-winning historical documentary *Spirits of Rebellion: Black Independent Cinema from Los Angeles* (2016) on the LA Rebellion film movement. His current book project for Indiana University Press is provisionally titled *Camerawork as Political Affect: Reenactment in Simulation Documentary*.

Index

Note: Page numbers followed by "n" denote endnotes.

For Product Safety Concerns and Information please contact our EU
representative GPSR@taylorandfrancis.com
Taylor & Francis Verlag GmbH, Kaufingerstraße 24, 80331 München, Germany

www.ingramcontent.com/pod-product-compliance
Lightning Source LLC
LaVergne TN
LVHW020642100826
845148LV00012B/2307

* 9 7 8 0 3 6 7 6 4 4 5 5 0 *